CMS Books in Mathematics
Ouvrages de mathématiques de la SMC

Nathalie Sinclair, David Pimm and
William Higginson (Eds.)

Mathematics
and the Aesthetic

New Approaches to an Ancient Affinity

 Springer

Nathalie Sinclair
Department of
  Mathematics
D216 Wells Hall
Michigan State University
East Lansing, MI 48824
USA
nathsinc@math.msu.edu

David Pimm
Department of Secondary
  Education
341 Education South
University of Alberta
Edmonton, AB T6G 2G5
CANADA
david.pimm@ualberta.ca

William Higginson
Faculty of Education
A235 McArthur Hall
Queen's University
Kingston, ON K7L 3N6
CANADA
higginsw@educ.queensu.ca

Editors-in-Chief
Rédacteurs-en-chef
K. Dilcher
K. Taylor
Department of Mathematics and Statistics
Dalhousie University
Halifax, Nova Scotia B3H 3J5
Canada
cbs-editors@cms.math.ca

Mathematics Subject Classification (2000): 00A30

Library of Congress Control Number: 2006926457

ISBN-10: 0-387-30526-2   e-ISBN -10: 0-387-38145-7
ISBN-13: 978-0387-30526-4   e-ISBN -13: 978-0387-38145-9

Printed on acid-free paper.

9 8 7 6 5 4 3 2 1

springer.com

This book is dedicated to the memory of
Martin Schiralli (1947–2003)
philosopher, colleague, teacher and friend.

PREFACE

A majority of the chapters in this book first saw the light of day as talks at a conference organised and held at Queen's University in Kingston, Ontario, Canada in April 2001. This small, invitational meeting, tellingly entitled *Beauty and the Mathematical Beast,* brought together a range of academics interested in and committed to exploring connections between mathematics and aesthetics. The enthusiastic response of participants at this gathering encouraged the presenters to expand upon their initial contributions and persuaded the organisers to recruit further chapters in order to bring a greater balance to the whole.

The timing of this event was not arbitrary. The preceding decade had seen a resurgence in serious writing dealing with deeper relations between mathematics (and science) and 'the beautiful'. In many ways, we the editors of this volume found these contributions to the literature were revisiting and drawing on themes that had been prominent over two thousand five hundred years ago, in certain writings of the Pythagoreans. While not intending to offer a historical reappraisal of these ancient thinkers here, we have none the less chosen to invoke this profound interweaving of the mathematical and the aesthetic to which this reputedly secretive philosophical sect was extensively attuned.

This book is divided into three sections comprising three chapters each, each with its own short introduction discussing the particular chapters within. These nine chapters in all are flanked by an introductory and a concluding chapter, both of which written by ourselves, which we describe now.

The opening Chapter α describes the ancient affinity between the mathematical and the aesthetic referred to in the book's title, an affinity we aim to illuminate as well as cultivate and advocate by means of this collection. Chapter α also provides a brief history of the mathematical aesthetic, beginning with the Pythagoreans but flowering in the twentieth century, while highlighting some of the themes and issues that subsequent chapters raise. These include attention to the following familiar questions: *can criteria for mathematical beauty be discovered?, is mathematics created or discovered?* and *is mathematics an art or a science?*

The final chapter of this book, Chapter ω, revisits some of these questions posed in Chapter α in light of the nine chapters in between. It provides some insights into those initial questions while raising further ones of its own. In particular, it offers three strong themes which stretch the mathematical aesthetic beyond the boundaries set by previous inquiries, all of which are related to potential sources of pleasure and desire for the mathematician: desire for distance and detachment; longing for certainty and perfection; pleasure in melancholy.

The ten authors of the various chapters in this book come from Canada, the US and Europe. Two who were born in Britain now live and work in Canada, while one from Latvia and one from Canada are now in the US. Each anglophone country has its own slight variants of spoken and written English, as well as punctuation conventions. Is the em-dash a thing of beauty or an abhorrence three times wider than any other character in the set? Is that extra 'u' in *colour* redundant, that repeated 'l' in 'travelled' an unnecessary extravagance (as a number of spell-checkers suggest)? Should the issue of the mathematical scope of variables enter into discussions of where to place commas and full-stops in relation to quotation marks? Is an 's' or a 'z' to be preferred in generalisations? [1] What seem to be matters of convention (and are therefore, at root, arbitrary) none the less raised a number of exercising aesthetic issues. As editors, we have decided on a position of plurality and respect for individual heritage, rather than impose a completely specified geographic orthography.

One of the considerable satisfactions we the editors have received in creating this book has arisen from drawing on the diverse expertise of the contributors to this volume, both mathematical and otherwise. Another has been the extended opportunity for the three of us to work alongside one another, exploring matters large and small.

We specifically want to mention here the breadth of scholarship that Martin Schiralli (the author of Chapter 5) brought to this project. Tragically, Martin died before this book was completed, aged only 56. His depth of philosophical knowledge, combined with his fresh perspective on mathematics, added considerably to many elements of this collection.

Nathalie Sinclair
David Pimm
William Higginson

January, 2006

[1] An entertaining discussion of some related issues can be found in *Eats, Shoots & Leaves* (Truss, 2004).

ACKNOWLEDGEMENTS

We, the editors, wish to acknowledge the considerable assistance provided by the combined resources of the following library systems:

Michigan State University
Queen's University
Simon Fraser University
University of Alberta
University of British Columbia
Vancouver Public Library

We also wish to thank Rachelle Painchaud-Nash (of Fine Line Design) for her meticulous and thoughtful design and setting of this book.

Chapter 3
Figure 1:   © David W. Henderson.
Figure 2:   © Daina Taimina.
Figure 10:   © 2004 The M.C. Escher Company – Baarn – Holland. All rights reserved.
Figure 18:   © 2004 The M.C. Escher Company – Baarn – Holland. All rights reserved.

Chapter 9
Figure 2:   Photographer W. Ritter, courtesy of Dumbarton Oaks, Byzantine Photograph and Fieldwork Archives, Washington, DC.
Figure 4:   © Photograph: The National Gallery, London.
Figure 5:   © Picasso Estate (Paris)/SODRAC (Montreal) 2005.
Figure 6:   © Museo Poldi Pezzoli, Milan.
Figure 7:   © René Magritte Estate/ADAGP (Paris)/SODRAC (Montreal) 2005 © Tate, London 2004.
Figure 8:   © Scala/Art Resource, NY.
p. 209:   Photograph by Ken Saunders.
p. 213:   Photograph by Graham Challifour.
Figure 9:   © The Trustees of the National Museums of Scotland.
p. 215:   Photograph © Lacock Abbey Publications, reproduced by permission of the William Henry Fox Talbot Trustees.
Figure 10:   © Réunion des Musées Nationaux/Art Resource, NY. Photograph by H. Lewandowski.

Chapter 10
Figure 1:   © The Trustees of The British Museum.

We are very grateful to the following for permission to reproduce poetic material in this book.

p. 45:   Jct Wimp (now Jet Foncannon), co-editor of the anthology *Against Infinity*, which contains the poem 'Poet as mathematician' by Lillian Morrison.

p. 182:   Sharon Nelson, for the lines quoted from her collection *This Flesh These Words*.

p. 226:   Patrick Lane, for the lines quoted from his collection *Old Mother*.

p. 248:   Faber and Faber Ltd (London) and HaperCollins (New York), for the lines quoted from Sylvia Plath's collection *Ariel*.

CONTENTS

NOTES ABOUT AUTHORS

Jonathan M. Borwein was the founding Director of the Centre for Experimental and Constructive Mathematics at Simon Fraser University (SFU). In 2004, he (re-)joined the Faculty of Computer Science at Dalhousie University as a Canada Research Chair in Distributed and Collaborative Research, cross-appointed in Mathematics, while preserving an adjunct appointment at Simon Fraser. Jonathan was born in St Andrews, Scotland, in 1951 and received his D.Phil. from Oxford in 1974 as a Rhodes Scholar. His interests span pure (analysis), applied (optimization) and computational (numerical and computational analysis) mathematics and high-performance computing. He has authored ten books (most recently two on experimental mathematics – www.expmath.info – and a monograph on variational analysis) and over 250 journal articles. He is currently Governor at large of the Mathematical Association of America (2004–07) and a past President of the Canadian Mathematical Society (2000–02).

David W. Henderson was born in Walla Walla, Washington State, and graduated from Ames (Iowa) High School, Swarthmore College (mathematics, physics, philosophy) and the University of Wisconsin (with a Ph.D. in geometric topology under R. H. Bing). After a two-year stint at the Institute for Advanced Study in Princeton, he joined the mathematics faculty at Cornell University in 1966 and has been there ever since. David's great love in mathematics is geometry of all sorts. His interests have widened into issues of what he calls educational mathematics. He has been directing Ph.D. theses in both mathematics and mathematics education. These interests led him to be invited to the ICMI Study Conference on the Teaching of Geometry in Sicily in 1995, where he met Daina Taimina. David has written four textbooks on geometry (three with Daina) and they have collaborated on many other activities. David has had visiting academic positions in India, Moscow, Warsaw, West Bank (Palestine), South Africa, USA and Latvia. Currently, he is Professor of Mathematics at Cornell University.

William C. Higginson is a member, and former Coordinator, of the Mathematics, Science and Technology Education Group in the Faculty of Education at Queen's University, Kingston, Ontario. A graduate of Queen's, Cambridge and Exeter Universities, as well as the University of Alberta, he has taught at Queen's since 1973. In 1983-84, he was a visiting professor in the Department of Architecture at the Massachusetts Institute of Technology and was a founding member of the Media Laboratory there. He returned to MIT as visiting professor of media technology in 1988. His research interests centre on the interaction between the subject of mathematics and various other disciplines, particularly literature as well as the visual and plastic arts. He is one of the authors of *Creative Mathematics* (Upitis, Phillips and Higginson, 1997), in which he documented a constructive–aesthetic approach to the teaching of mathematics.

R. Nicholas Jackiw is the original designer and developer of *The Geometer's Sketchpad*. An educational software environment for the creation, visualization, exploration, and analysis of mathematical models, Sketchpad provides a rare example of effective educational software that has made a successful transition from academic research lab to widespread commercial impact. He began work on the software as an undergraduate at Swarthmore College, under the direction of Eugene Klotz and Doris Schattschneider, and presently serves as the Chief Technology Officer of KCP Technologies, the software affiliate of Key Curriculum Press. In addition to designing software and directing the project's programming staff, he is active in the pre-service and in-service professional development of teachers, conducting workshops and institutes across the country; and has overseen the program's translation into more than a dozen foreign languages. He leads NSF software research projects in Principal Investigator or Senior Researcher capacities and is a frequently invited speaker to schools, software and geometry conferences, as well as meetings of the National Council of Teachers of Mathematics.

David Pimm is currently a professor of mathematics education at the University of Alberta. A Fulbright scholar, he has authored three books, edited four more and written many research articles which explore the inter-relationships between language and mathematics. His work has focused both on analyses of mathematics classroom language and on producing theoretical accounts of linguistic aspects of mathematics itself. He is particularly interested in the roles of metaphor and metonymy in creative mathematical endeavour. His secondary research interest is in the potential influence of studies of the history and philosophy of mathematics on the teaching of mathematics. He was editor of the international journal *For the Learning of Mathematics* from 1997 until 2003.

Doris Schattschneider holds a Ph.D. in mathematics from Yale University and is Professor Emerita of Mathematics at Moravian College, where she taught since 1968. She was Geometer and Senior Associate on the NSF-funded Visual Geometry Project that produced the software *The Geometer's Sketchpad,* along with videos and activity books on polyhedra and symmetry. She has lectured widely on the topics of tiling, polyhedra, symmetry, dynamic geometry, geometry and art (especially the art of M. C. Escher) and visualization in teaching. She is the author of more than 40 articles, and author, co-author or editor of several books, including *M. C. Escher: Visions of Symmetry* (Freeman, 1990; Abrams, 2004), *Geometry Turned On: Dynamic*

Software in Learning, Teaching, and Research (MAA, 1997), *A Companion to Calculus* (Brooks/Cole, 1995) and *M. C. Escher's Legacy* (Springer, 2003). Doris has been active in the Mathematical Association of America at all levels and was editor of *Mathematics Magazine* 1981–1985. In 1993, she received the national MAA Award for Distinguished Teaching of College or University Mathematics.

Martin Schiralli, until his untimely death in 2003, was associate professor of the philosophy of education at Queen's University where he specialised in epistemology and aesthetics. His book, *Constructive Postmodernism* (1999), analysed the epistemological challenges and aesthetic opportunities presented in newer ways of thinking. In keeping with the renewed postmodern interest in relaxing the epistemological demarcations between established disciplines and fields of inquiry, his most recent research involved the concept of 'pattern' as a means of representing those underlying affinities between the mathematical and the aesthetic that are of particular relevance to contemporary art and technology.

Nathalie Sinclair is currently an Assistant Professor at Michigan State University, cross-appointed between the Department of Mathematics and the Department of Teacher Education. Her work on the mathematical aesthetic has extended to the K-12 domain, and, more recently, to the post-secondary level where she is investigating the aesthetic development of young mathematicians. She is also interested in dynamic mathematics environments and the roles of visualisation, intuition and experimentation in the development of mathematical understanding.

Dick Tahta has taught both in schools and universities. Now retired, he still has a love–hate relationship with mathematics. He often feels iconoclastic about mathematics – especially the teaching of it to the unwilling. But sometimes a theorem can excite him as much as a painting or a poem.

Daina Taimina was born and received all her formal education in Riga, Latvia. In 1977, she started to teach at the University of Latvia and continued for more than twenty years. Her Ph.D. thesis was in theoretical computer science (under Rusins Freivalds), but later she became more involved with geometry, history of mathematics and mathematics education. These interests led to her being invited to the ICMI Study Conference on the Teaching of Geometry in Sicily in 1995 where she met David Henderson. She has written a book on the history of mathematics (in Latvian) and (with David) three recent geometry textbooks. Daina was a Visiting Associate Professor at Cornell 1997-2003. Currently, Daina is a Senior Research Associate at Cornell University.

CHAPTER α

A Historical Gaze at the Mathematical Aesthetic

Nathalie Sinclair and David Pimm

> No matter how correct a mathematical theorem may appear to be,
> one ought never to be satisfied that there was not something imper-
> fect about it until it gives the impression of also being beautiful.
> (George Boole, in MacHale, 1993, p. 107)

The ancient Greeks, primarily by way of the Pythagoreans, established and celebrated a fundamental affinity between the mathematical and the aesthetic. This affinity was nothing about surface charm or happy coincidences. It had deep roots, integral as it was to the world-view of the Pythagoreans, to their beliefs about knowledge and learning. It closely connected the raw world of sense and experience to the divine world of perfection and beauty. Number was the principle that governed all things, rather than being simply useful for counting or measuring – as modern minds might think, if indeed they stop to consider this omnipresent convenience at all. Through number, one could come to know the world, and through the harmonies found in numerical patterns and in geometrical forms, one could gain access to the clearest and most indubitable essence – the real.

This ancient affinity started losing sway early on, even with Plato and Aristotle. Nevertheless, traces of this Pythagorean perception have remained, resurfacing at various times, such as at the beginning of the twentieth century. For instance, in the second volume of his book *On Growth and Form,* D'Arcy Thompson (1917/1968) wrote:

> For the harmony of the world is manifest in Form and Number,
> and the heart and soul and all the poetry of Natural Philosophy are
> embodied in the concept of mathematical beauty. (pp. 1096-1097)

Thompson went on to add that this is what the Pythagoreans taught us, Philolaus in particular (a Pythagorean whose influence is also discussed in Chapter 5 of this book), before remarking:

> Moreover, the perfection of mathematical beauty is such (as Colin
> Maclaurin learned of the bee), that whatsoever is most beautiful
> and regular is also found to be most useful and excellent. (p. 1097)

While not all would agree with his attributing to the beautiful the most utilitarian properties (or at least the most 'useful', however that may be seen at different times), Thompson is, in this passage, apparently identifying mathematics as possessing the highest form of perfection – a theme we shall find recurring repeatedly. Finally, Thompson seconded the view of a certain Monsieur Henri Fabré, who wrote that one sees in Number "le comment et le pourquoi des choses" [the how and the why of things] and finds in it "la clef de voûte de l'Univers" [the keystone of the universe] (p. 1097).

If Thompson signalled rapprochement, we can also identify periods of disjunction or even denial of this ancient affinity. For instance, T. S. Eliot (1921/1932) wrote of his sense of a 'dissociation of sensibility' (the loss of the direct fusion of thought and feeling) in much of the poetry of eighteenth- and nineteenth-century England. (This, as well as other instances, including the nineteenth-century English Romantics' scorn of mathematics and science, is touched on in Chapter 9.)

And one of the more recent accounts of this process of the scientific/ artistic affinity dissolving (at least symbolically), from the mid-twentieth century this time, was given by scientist and novelist C. P. Snow (1959) in his essay naming and exploring aspects of 'the two cultures'. Since then, however, these two cultures – the arts and the sciences – have once again started to find an intermittent, yet growing rapport, as evidenced by the number of books, conferences and courses seeking common behaviour and beliefs.

This recent work has included many productive marriages between the sciences (including mathematics) and the arts, such as, for example, contemporary sculptures of numerical patterns (Dickson, 1993) and mathematical analyses of Jackson Pollock's paintings (Taylor *et al.*, 1999): this is further discussed in Chapter 6. Ethnographically-oriented scholars have taken interest in revealing the mathematical dimension of past artistic artifacts, such as the geometry of Pueblo pottery (Campbell, 1989) and the symmetry of Islamic design (Chorbachi, 1989). And, of course, the plethora of books on the Dutch artist Maurits Escher, particularly the recently published *M.C. Escher's Legacy: a Centennial Celebration* (Schattschneider and Emmer, 2003), has shown how his prints were born out of the artist's mathematical *and* artistic interests and how his work continues to inspire both mathematical *and* artistic analyses.

Scholars working in this interdisciplinary, 'cross-cultural' arena provide concrete examples of the ways in which mathematics and the arts can both inspire each other, not only in contemporary settings, but also in historical ones. Increasingly, however, scholars have also been working to reveal the close relationship between scientific and artistic creativity and have succeeded in defying popular beliefs that feed the antagonistic 'two culture' worldview, including that which holds that scientists operate exclusively rationally and artists solely intuitively or emotionally.

Some aspects of this ancient affinity even seem to be seeping into other, non-academic cultural milieux. Images of fractals, for instance, which have become increasingly widespread (who has never found themselves staring at a fractal screensaver?), have provided many non-mathematicians with opportunities to encounter compelling, visually beautiful mathematical artifacts.

And though mathematics is far from being seen as playing the central epistemological role it did for the Pythagoreans, it has nonetheless made some inroads into more mainstream culture.

The proliferation of mathematical films and plays, such as *A Beautiful Mind, Pi, Arcadia, Breaking the Code* and *Proof,* harken back to ancient times when playwrights such as Aristophanes could refer to then-current mathematical problems (such as the squaring of the circle) as easily as political ones. Publishers have apparently realised that the once-sullen, esoteric line of pure mathematics books might be gaining in appeal, as titles such as *Fermat's Last Theorem, The Code Book* and *The Honor Roll: Hilbert's Problems and their Solvers* populate bookstore shelves. Instead of offering accounts of mathematics using the formal, abstract language to be found in research journals – and often imposed upon reluctant schoolchildren – these books tell exciting, sometimes heart-wrenching and very human stories of mathematicians and their discoveries, seeking to convey the sense of beauty and elegance to which mathematicians are drawn. Once again, we are being provided with glimpses of the way in which mathematics connects experience and abstraction, connects the senses with structures, connects the human with the divine.

The scale of the recent rapprochement among mathematics, science and the arts, as well as the apparently growing appeal of mathematics in more mainstream culture, are both manifestations of a re-emergent affinity between the mathematical and the aesthetic, one that might be coming closest to the golden era of the Pythagoreans. In keeping with the philosophy of the Pythagoreans, the chapters in this book focus on this affinity at a deeper level, beyond surface applications (as might be suggested by geometricised paintings or musical fractals), to more fundamental, epistemological connections. They attempt to articulate a common sub-stratum between the mathematical and the aesthetic, one that is integrally related to human sense-making and to learning.

The goal of this opening chapter is to provide a brief historical sense of the development of ideas around the mathematical aesthetic. Readers with backgrounds in the aesthetics branch of classical philosophy will find the equivalent branch of mathematics rather young and comparatively uncritical. Nevertheless, a certain amount of grappling with difficult challenges can be found, though not with the systematic or cumulative attention that has built and continues to build the mathematical edifices so cherished by mathematicians themselves.

We begin by looking at some fragments of these challenges, as found in the long period stretching from the ancient Greeks up to the beginning of the twentieth century. We then turn to the twentieth century itself and find there an explosion of interest in the mathematical aesthetic, particularly around questions such as: *is mathematics an art or a science?* and *can criteria for mathematical beauty be identified?*

Some Pre-Twentieth-Century Fragments
Concerning the Mathematical Aesthetic

The extant writings attributed to Pythagoras and his followers reveal that the Pythagorean school, if not Pythagoras himself, found in the beauty of mathematics the very highest order of aesthetic interest. In fact, the Pythagoreans were overwhelmed by the aesthetic appeal of the theorems they discovered and were perennially preoccupied with the interconnectedness of the mathematical and the aesthetic. This interconnectedness permeated their worldview, which saw reality as ultimately revealed in mathematically harmonious concepts.

Mathematical studies were thus seen as furnishing ladders and bridges to the divine, because they shared a perfection and beauty that was considered true of the divine, but felt lacking in the physical world. Unlike Plato, who separated number, an abstract entity, from the things numbered, the Pythagoreans saw number as being tied up with the actual procedure of counting and thus closely connected with things. Number reached out or down into the world of sense and experience. As such, the Pythagoreans saw the roots of Plato's *exclusively* abstract entities in the 'real', the human, the sensory world.

Both Plato and Aristotle, though philosophically divergent in many ways, were much influenced by the ideas of Pythagoras, particularly with respect to the connection between mathematics and the beautiful. Plato saw mathematics as providing the most fundamental of all ideas and believed in mathematical objects as perfect forms. As he wrote in *Philebus*:

> By 'beauty of figures' I mean in this context not what most would consider beautiful – not, that is, the figures of creatures in real life or in pictures. I mean a straight line, a curve and the plane and solid figures that lathes, rulers and squares can make from them. I hope you understand. I mean that, unlike other things, they are not *relatively* beautiful: their nature is to be beautiful in any situation, just as they are, and to have their own special pleasantness, which is utterly dissimilar to the pleasantness of scratching. (51d; 1982, p. 121; *italics in Waterfield*)

And Aristotle, in his *Metaphysics,* wrote that the mathematical sciences have much to say about the beautiful and the good, and that:

> the chief forms of beauty are order and symmetry and definiteness, which the mathematical sciences demonstrate in a special degree. (Book M, 1078b; 1966, p. 218)

(Martin Schiralli, in Chapter 5, however, discusses important differences among the views of Plato, Aristotle and the Pythagoreans.)

Once into the Christian era, by no means all were comfortable with linking the mathematical and the divine, of humans equating investing the mathematical and investing the divine with the qualities of perfection, thereby perhaps equating the two. For instance, St Augustine, in his twelve-volume work *De Genesi ad Litteram,* warned:

> Hence, the good Christian should beware of mathematicians and all those who make empty prophecies, especially when they tell the truth, for fear of leading his soul into error by consorting with demons. (Book II, *23*, 35-36)

When reading this observation, however, it is important to realise that the most common connotative meaning of the word 'mathematician' in St Augustine's day was not what it would be today, including as it did those engaged in astrology, alchemy, *gematria* and magic. And, as Chapter 9 speculatively explores, in Byzantium and in mediaeval Europe at least, the drive to mathematise may have been 'side-tracked' into theology, until Renaissance artists found an alternative outlet in their work.

With regard to Islam, Endress (2003) informs us that the only mediaeval mathematics-related dissertation on the aesthetically beautiful can be found in Ibn al-Haytham's *Optics,* a discipline that was seen as the converse of geometry by mediaeval mathematicians. It may be true that such mathematicians were less inclined to talk directly about the beauty of mathematics; nonetheless, they certainly wrote about some of its other aesthetic qualities. For example, the tenth-century mathematician Abu Salh al-Kuhi – according to Berggren perhaps "the last mathematician to look on mathematics with the eyes of the great Hellenistic geometry" (cited in Endress, p. 193) – extolled the certainty of mathematics. He wrote of the rules of geometry as being "consistent and unchanging" and eschewed the kind of 'bad' mathematics that was based on numerical, imperfect approximations.

The eleventh-century Islamic theologian Al-Ghazzali warned of mathematics – and particularly its predilection for aesthetic qualities such as precision and clarity – leading to harmful things other than magic. One additional drawback of mathematics, he wrote, was that:

> every student of mathematics admires its precision and the clarity of its demonstrations. This leads him to believe in the philosophers and to think that all the sciences resemble this one in clarity and demonstrative power. (in Hoodbhoy, 1991, p. 105)

Such caveats against misplaced or even idolatrous authority, whether to be located in a particular author or within mathematics itself, have been echoed time and time again down the centuries. For instance, we note in passing that one of the more notable complaints concerning Isaac Newton's extensive biblical chronology (which occupied much of the latter part of his life) was its being credited with more credence than its due, because of the reputation of its creator in quite another area of human endeavour.

So, over a millenium after St Augustine's expression of concern, we still find Archbishop François de Fénélon (1697/1845) in Paris expressing a not-dissimilar unease:

> Surtout ne vous laissez point ensorceler par les attraits diaboliques
> de la géométrie. [Above all, do not allow yourself to be bewitched
> by the diabolical attractions of geometry.] (p. 493)

In the Christian West, right up to the time of Fibonacci (and beyond, into
the sixteenth and even seventeenth centuries), the more likely meaning for
'mathematician' was astrologer (and, even worse, 'conjuror'). It is worth
recalling that such an Augustinian pejorative description of 'mathematician'
(or its common equivalent of 'geometer') was almost as fitting of Isaac
Newton (see, for instance, Gleick, 2003, on the 'alternative' Newton) as the
Elizabethan neoplatonist mathematician and magus John Dee (1527–1608)
of an England a century earlier, whose magnificent personal academic library
was perhaps the best in England at that time (see Yates, 1969).

Dee lived in very complex political, religious and intellectual times.
Similar concerns linking mathematics with devil-worship surfaced in England,
very soon after the English Reformation started, with Henry VIII asserting
the King as head of the new Church of England (via the 1534 Act of
Supremacy denying the authority of the Pope). In 1550, three years after the
death of Henry VIII, government commissioners ('Visitors') went destruc-
tively through Oxford University college libraries, casting more than a sus-
picious glance at books containing mathematical diagrams, consigning many
volumes to destruction. [1]

Twenty years after this book-burning event, Dee published his extensive and very influential 'fruitfull præface' (which ran to ninety-five printed
pages) to the first English-language version of Euclid's *Elements*. Following
a highly Pythagorean discussion of the nature of mathematics in terms of
number, Dee asserted:

> For, [*Things Mathematicall*], being (in a manner) middle, between
> things supernaturall and naturall: are not so absolute and excellent
> as things supernaturall; Nor yet so base and grosse, as things nat-
> urall: But are things immateriall, and neverthelesse, by material
> things able somewhat to be signified. And though their perticular
> Images, by Art, are aggregable and divisible: yet the generall *Forms*
> notwithstanding, are constant, unchangeable, untransformable and
> incorruptible. Neither of the sense, can they, at any time, be per-
> ceived or judged. Nor yet, for all that in the royall mind of man,
> first conceived. But surmounting the imperfection of conjecture,
> weening and opinion: and comming short of high intellectuall con-
> ception, are the Mercuriall fruit of *Dianœticall* discourse, in per-
> fect imagination subsisting. A marvellous newtrality have these
> things *Mathematicall,* and also a strange participation between
> things supernaturall, immortall, intellectuall, simple and indivisible:
> and things naturall, mortall, sensible, compounded and divisible.
> Probability and sensible proof, may well serve things naturall, and
> is commendable: In Mathematicall reasonings, a probable Argu-
> ment is nothing regarded: nor yet the testimony of sens[e], any
> whit credited: But onely a perfect demonstration, of truths certain,

necessary, and invincible: universally and necessarily concluded: is allowed as sufficient for an Argument exactly and purely Mathematicall. (1570; in Rudd, 1651, pp. 4-5)

There are a number of resonances between the above quotation of Dee's and themes addressed in this book. First, in placing mathematics neither of this world nor the next, but somehow hovering between the two with connections and links to both, Dee calls attention to the Janus-faced nature of mathematics, as well as presciently identifying mathematics as a 'mediating third' between the two.

To evoke in the context of this quotation the tension between 'pure' and 'applied' mathematics, to cast it in this modern frame (that is, to worry about Eugene Wigner's (1960) claim about 'the unreasonable effectiveness of mathematics'), is to assert the gap between mathematical and natural. But some of the protestors cited above are at equally great pains to maintain the separation between mathematical and what Dee terms 'the supernaturall', identified by some (but not all) with 'the divine'.

We would also like to draw on this quotation in order to make some links with themes explored in this book. To a considerable extent, quite a number of chapters in this book – in particular, Chapters 3, 4, 5, 8 and 9 – explore different ways of disagreeing with Dee's remark "Neither of the sense, can they, at any time, be perceived or judged". Additionally, in Chapter 1, Jonathan Borwein takes (indirect) exception to Dee's assertion that "a probable Argument is nothing regarded". David Pimm, in Chapter 8, discusses aspects of what Dee termed the "Art" of "perticular Images", as well as exploring the connection between 'Popish' catholicism and concern about mathematical images in the twentieth century (prefaced, as we saw above, in the sixteenth). Finally, in Chapter 9, Dick Tahta centrally examines the nature of "sensible" objects in relation to mathematics.

The Mathematical Aesthetic in the Twentieth Century

Though the eighteenth and nineteenth centuries were extremely fruitful in terms of mathematical discoveries and advances, it seems that mathematicians infrequently, at least in print, reflected on issues related to the mathematical aesthetic. This is not to say, however, that they did not think about or mention aesthetic values. Gauss's mathematical diary (see Gray, 1984), for example, contains many references to the beauty or elegance of his own mathematical ideas and discoveries. For instance, as a nineteen-year-old in 1796, Gauss wrote about a new proof obtained "all at once, from scratch, different, and not a little elegant" (p. 108). In another entry, this time made in 1800, he described his work on the arithmetic–geometric means as being "most beautifully bound together and increased infinitely" (p. 122) to the theory of transcendental quantities. In addition to beauty and elegance, Gauss made reference to aesthetic qualities such as a "charming theorem" (p. 125) and to a "most simple and expeditious method" (p. 124).

However, for some reason, the turn of the past century brought about a comparative flurry of interest in the nature of mathematics. In particular, there were concerted efforts to ascertain whether mathematics belonged more to the arts or to the sciences, from which it had not long ago been divorced (during the latter part of the nineteenth century, not least due to developments in connection with non-Euclidean geometry). It also marked the beginning of sustained inquiries into the development of mathematical knowledge and the extent to which it is fuelled by some aesthetic as well as utilitarian or logical considerations (which, *pace* D'Arcy Thompson, were usually seen as relatively distinct).

Finally, and early on in this flurry of activity, mathematicians became interested once more in the psychology of mathematical discovery. [2] Some twentieth-century mathematical writers on the aesthetic turned to the central question of the extent to which affective responses and aesthetic sensibilities were involved in the process of mathematical creation. Their attention to the aesthetic was not as intense and all-encompassing as that of the earlier Pythagoreans, but they each began, in their own way, to rekindle the embers of this ancient affinity. Here, we examine each of these themes in turn, tracing out, when possible, aspects of their historical developments.

The aesthetics of mathematical creation

In 1908, Henri Poincaré began to bring renewed attention to the aesthetic dimension of mathematical creation, but his focus was more pragmatic and markedly different from that of the ancient Greeks. He was most interested in probing the aesthetic influences that affect the process of mathematical discovery. This focus proved unlike that of many of the mathematicians who would follow him, who attended more to the aesthetic values or principles that exist in mathematical ideas or products (the discoveries themselves). By analysing the process of mathematical creation, Poincaré tried to show that mathematical invention depends upon the often sub-conscious choice and selection of 'beautiful' combinations of ideas, those best able to "charm this special sensibility that all mathematicians know" (1908/1956, p. 2048).

In his book *The Psychology of Invention in the Mathematical Field,* Jacques Hadamard (1945) proposed the first expansion of Poincaré's aesthetic heuristic theory, additionally claiming that aesthetic sensibilities often guide a mathematician's general choices about which line of investigation to pursue. He wrote specifically about the "sense of beauty" (p. 130) which can inform the mathematician that "such a direction of investigation is worth following; we feel that the question *in itself* deserves interest" (p. 127; *italics in original*). Hadamard also added to Poincaré's ideas on the role of the mathematical pre-conscious in mathematical thinking, locating the period in which it is most operative – the *incubation* period – within a larger theory of mathematical inquiry. Through both historical and empirical studies, he supported his account from mathematicians such as Pierre de Fermat, Evariste Galois, Bernhard Riemann, George Birkhoff, George Pólya and Norbert Wiener.

Morris Kline (1953) subsequently pointed out that aesthetic concerns not only guide the direction of an investigation, but motivate the search for new proofs of theorems already correctly established but lacking in aesthetic appeal – by means of their ability to "woo and charm the intellect" (p. 470) of the mathematician. Kline took this aesthetic motivation as a definitive sign of the artistic nature of mathematics. Wolfgang Krull (1930/1987) illustrated how aesthetic preferences – such as a mathematician's desire for simple, symmetric structures – can seriously influence the further development of mathematics, as well as the derivation of new properties and the creation of new theories.

In his earlier attempt to define mathematics as the "classification and study of all possible patterns" (p. 12), Warwick Sawyer (1955) implied that the heuristic value of mathematical beauty stems from mathematicians' sensitivity to pattern and originates in their belief that *where there is pattern there is significance"* (p. 36; *italics in original*). Sawyer went on to explain the heuristic value of this trust in pattern:

> If in a mathematical work of any kind we find that a certain striking pattern recurs, it is always suggested that we should investigate *why* it occurs. It is bound to have some meaning, which we can grasp as an idea rather than as a collection of symbols. (p. 36; *italics in original*)

Sawyer might well have explained Poincaré's special aesthetic sensibility as a sensibility toward pattern, viewed broadly as any regularity that can be recognised by the mind. For him, the mathematician is not only able to recognise regularities and symmetries, but is also attuned to look for and respond to them with further investigation.

Poincaré's writing on the mathematical aesthetic, which was definitely excluding of most everyone (more so than Sawyer's account) suggested that only the very creative mathematicians had access to this aesthetic guide. This claim may have provoked the "literary superstition" that Alfred North Whitehead (1926) mentioned, which views the aesthetic appreciation of mathematics as being a "monomania confined to a few eccentrics in each generation" (in Hardy, 1940, p. 85). Hardy quoted Lancelot Hogben (1940) "the aesthetic appeal of mathematics may be very real for a chosen few" (p. 86) and accused him of echoing this "superstition".

Indeed, Bertrand Russell's (1917) famous quotation, "Mathematics, rightly viewed, posseses not only truth, but supreme beauty – a beauty cold and austere, like that of sculpture" (p. 57), does seem to suggest that mathematics exercises a coldly impersonal attraction, one not meant for normal individuals. As we shall see, Russell's frigid tastes are not the only ones that mathematics can satisfy. But this theme of the exclusiveness of mathematical aesthetic judgements (concerning who is able to make them), to be found in the writings of Poincaré, Russell and Hardy, persists in the mathematics literature.

Armand Borel (1983) was faced with overcoming a different kind of exclusivity in his attempt to convey the nature of mathematics and the mathematical aesthetic to a wider audience, of both mathematicians and non-mathematicians. He began by arguing that the development of mathematics was "derived from, guided by, and judged according to aesthetic criteria" (p. 11), thereby astutely acknowledging both Poincaré's heuristic aesthetic and Hadamard's aesthetic of choice. However, he then attempted to show how what may seem like the "pure and esoteric" aesthetics of mathematicians are actually bound up with "more earthly yardsticks" (p. 15), such as applicability and usefulness, values that Borel hoped non-mathematicians would find more recognisably mathematical than beauty or elegance.

Almost eighty years after Poincaré, the philosopher Harold Osborne (1984) wrote:

> the reliance on the heuristic value of mathematical beauty in scientific theory has become something of a commonplace. (p. 291)

This indicates the extent to which scientists – and especially physicists – had placed their trust in Poincaré's notion of the mathematical aesthetic sensibility as a kind of muse who, if listened to carefully, would both guide and inspire creativity. [3] Indeed, scientists have been much more prolific than mathematicians in cataloguing and inspecting the effect of this trust on the development of scientific theories (see, for example, Chandrasekhar, 1987; Curtin, 1982; Farmelo, 2002; McAllister, 1996; Wechsler, 1978).

Yet few scholars have explicitly discussed the *differences,* in terms of their aesthetic dimensions, between mathematics and the (other) sciences. There is certainly a common belief among physicists that what they find beautiful in their theories is ultimately mathematical. In fact, it would seem that mathematics plays a key bridging role between the sciences and the arts, at times transforming scientific ideas into forms and patterns that afford aesthetic attention. But science and mathematics have different aims, as well as different measures of success. This leads McAllister (1996), for instance, to warn that the nature and role of the scientific aesthetic cannot be blindly transferred to the domain of mathematics.

Mathematics: an art or a science?

The mathematics literature has long been replete with questions about the nature of mathematics and its place in the plural world of the arts and sciences. While Gauss's claim that mathematics is the queen of the sciences has often been repeated, so has the claim that mathematics belongs more properly to the arts. The British scholar J. W. N. Sullivan made the latter argument in 1925, claiming that mathematics is the product of a free creative imagination, unconditioned by the external world. It is, he argued, just as 'subjective' as the other arts, even though it can be used to illuminate natural phenomena.

Moreover, Sullivan (1925/1956, p. 2020) claimed that mathematicians are impelled by the same incentives as artists, citing as evidence the fact that the "literature of mathematics is full of aesthetic terms" and that many mathematicians are "less interested in results than in the beauty of the methods" (p. 2020) by which those results are found. His interest in the mathematical aesthetic experience, which he saw giving rise to the same satisfactions as the artistic experience, was distinct from Poincaré's focus on the mathematical aesthetic sensibility, which acts as a guide. Yet Sullivan saw neither mathematics nor art as existing to satisfy "aesthetic emotions": rather, he saw both art and mathematics as means by which humans can "rise to a complete self-consciousness" (p. 2021).

The philosopher Rom Harré (1958) was more interested in the aesthetic *differences* between mathematics and the arts. He pointed out the uniqueness of mathematical aesthetic judgements by comparing them with *bona fide* aesthetic judgements. He described mathematical appraisals of beauty and elegance as *quasi-aesthetic,* since they use "words from our regular aesthetic vocabulary, which fall outside the normal range of aesthetic judgements" (p. 133). In fact, for Harré, "quasi-aesthetic appraisals are not a queer sort of aesthetic appraisal but simply not aesthetic appraisals at all" (p. 136). Quasi-aesthetic appraisals are "essentially second-order" because of two factors (p. 137). First, appraisals such as 'beautiful' and 'elegant' do not betoken success in mathematics the way they do in artistic fields: "If an object doesn't move us it has failed altogether aesthetically, but if a proof doesn't move us it does not for that reason fail altogether mathematically" (p. 137). Second, quasi-aesthetic appraisals require *comparing* an object with very specific other objects of the same kind. Harré contended that, in mathematics, the elegance of a proof "can only be judged by a comparison, explicit or implicit, between alternative proofs of the *same* result" (p. 137): in contrast, "Ordinary aesthetic appraisals are essentially non-comparative" (p. 137).

Harré's formalist stance forced him to trivialise almost completely the importance of aesthetic appraisals in mathematics. Because the aesthetic is only secondary to achievement, it is thereby robbed of any epistemic interest. Furthermore, contemporary philosophers and art theorists would challenge Harré's claim about the aesthetic bar of success inherent in the arts and the non-comparability of aesthetic judgements. The philosopher of mathematics Thomas Tymoczko (1993) may well have pointed out the most operative difference between aesthetic judgements in mathematics and those at work in the arts. This is that the mathematics community does not have many (any?) 'mathematics critics' to parallel the strong role played by art critics in appreciating, interpreting and arguing about the aesthetic merit of artistic products.

In 1933, the American mathematician George Birkhoff approached the connection between mathematics and beauty from the reverse direction, proposing a theory by which mathematics could be used to describe beauty. According to Whittaker (1945), Birkhoff wanted to create:

> a general mathematical theory of the fine arts, which would do for aesthetics what had been achieved in another philosophical subject, logic, by the symbolisms of Boole, Peano, and Russell. (p. 127)

Birkhoff admitted that the aesthetic feeling was "intuitive" and *"sui generis"*, but held nevertheless that the attributes upon which aesthetic values depend are accessible to measurement. He proposed three main variables constituting typical aesthetic experiences: the complexity of the object (C), the feeling of value or aesthetic measure (M) and the property of harmony, symmetry or order (O). With the following equation, $M = O/C$, he presented to us his hypothesis that the aesthetic measure is determined "by the density of order relations in the aesthetic object" (1933/1956, p. 2186). He also provided equations that could define both the variables O and C more formally.

Birkhoff's formula never gained much currency in the world of art criticism, nor in the world of mathematics. After all, the terms O and C are not straightforward to measure: can the square grid, which is highly ordered with little complexity, be considered of great aesthetic value? What about a fractal image? The difficulty in measuring O and C makes the formula almost impossible to use. And perhaps artists and mathematicians alike were unimpressed by Birkhoff's formula for its tacit presumption that aesthetic value can be measured in some absolute way (regardless of personal, social or cultural styles), based on a set of accurate rules. Regardless of his formulaic approach, Birkhoff did identify qualities such as order, harmony and complexity as being relevant to aesthetic value, thus echoing the ancient Greeks while at the same time anticipating the work of several of the mathematicians we have yet to discuss.

Criteria for the mathematically beautiful

In 1940, G. H. Hardy published what became arguably the most widely-read inquiry into the mathematical aesthetic. Unlike either Poincaré or Hadamard, Hardy was primarily interested in defining mathematical beauty as it exists in mathematical products, particularly in proofs. He proposed a somewhat complex scheme that distinguished 'trivial' beauty – which can be found in chess – from 'important' beauty, which can only be found in serious mathematics. But, for Hardy, serious involved significant, which in turn required generality – scope or reach – and depth. *Generality* and *depth* are both difficult to define, but can, according to Hardy, be immediately recognised by those with a "high degree of mathematical sophistication" (p. 103). Such mathematicians will find a mathematical idea significant when it can be "connected, in a natural and illuminating way, with a large complex of other mathematical ideas" (p. 89). Hardy illustrated his notion of mathematical beauty with two examples: Euclid's proof of the infinity of primes and the Pythagorean proof of the irrationality of $\sqrt{2}$. These two proofs appear frequently in the literature as particularly fine examples of beautiful proofs (for example, see Dreyfus and Eisenberg, 1986, or King, 1992).

Chapter α – A Historical Gaze at the Mathematical Aesthetic   13

Having defined mathematical beauty in terms of significance and seriousness, Hardy went on to say that the triviality of ideas (such as those found in chess problems, but not in beautiful mathematics) "disturbs any more purely aesthetic judgement" (p. 113). Hardy proposed that purely aesthetic qualities are unexpectedness, inevitability and economy. Considerably later, Roger Penrose (1974) would add to Hardy's list the criterion of "unexpected simplicity" (p. 267). Hardy advanced a formalist perspective of mathematical beauty by only acknowledging responses to formal properties. For Hardy, and many others, formalism represents the dominant 'public aesthetic' of mathematics; if mathematics presents *any* aesthetically relevant qualities, these qualities *must* be formal in nature.

Shortly after Hardy's publication, François Le Lionnais (1948/1971) proposed a completely different, non-formalist way of approaching the problem of mathematical beauty – without making reference to either Hardy or Poincaré. Le Lionnais was not interested in the process-oriented aesthetic sensibilities that Poincaré was, but his scope was wider than Hardy's, including as it did various kinds of 'facts' and 'methods' as potential objects of mathematical beauty. Le Lionnais effectively enlarged the sphere of mathematical entities that can have aesthetic appeal, including not only entities such as definitions, shapes, proofs, solutions and theorems, which are appreciated after the fact, but also the various methods and processes used to work with mathematical entities, which can be appreciated *while* doing mathematics.

In addition, Le Lionnais emphatically drew attention to the subjectivity of aesthetic responses, by classifying mathematicians' orientations as either 'classical' or 'romantic', thus allowing for degrees of appreciation – banned by Hardy – according to personal preference. These categories represent two styles of human endeavour: on the one hand, a desire for equilibrium, harmony and order; and, on the other, a yearning for lack of balance, form obliteration and pathology. A very similar distinction was made by Freeman Dyson (1982), who distinguished between 'unifiers' and 'diversifiers', the former finding and cherishing symmetries, the latter enjoying the breaking of them.

In addition, Harold Osborne (1984), in tracing aspects of the aesthetic in the sciences, also recognised the human dimension of mathematical aesthetic response, arguing that aesthetic satisfaction derives from the common human desire to impose order on chaos. Citing Davis and Hersh's (1980) observation, "to some extent, the whole object of mathematics is to create order where previously chaos seemed to reign, to extract structure and invariance from the midst of disarray and turmoil" (p. 172), Osborne implied that mathematics provides an optimal context in which to gain aesthetic satisfaction.

Le Lionnais's stance on the subjectivity of aesthetic responses did not do much to quell the belief, common among mathematicians especially, that most mathematicians will agree on their aesthetic judgements. This common belief was fuelled in part by the exclusivity of Poincaré and Hardy, which seemed to imply that if your aesthetic judgement did not agree with that of a great mathematician, then you were simply not a great mathematician. It was also fuelled by the enormous discrepancies of taste and judgement found in the arts which, by any mathematician's definition of subjectivity, dwarfed the differences identified in the mathematical world.

Jerry King (1992), like Hardy, presumed the supposed homogeneity of mathematicians' aesthetic response and further concluded that mathematicians work from some set of commonly-accepted aesthetic principles. Moreover, he assumed that mathematicians' judgements are not subjective, but instead depend solely upon the mathematics itself, making it possible to formulate decisive criteria. In his book *The Art of Mathematics,* King drew on aesthetic theories of philosophy and art criticism in order to articulate "a complete aesthetic theory of mathematics" (p. 157).

Rather than expanding Hardy's or Osborne's list of factors that contribute to aesthetic appeal, King's primary goal was to identify general-level aesthetic criteria that would help distinguish 'good' mathematics from 'bad' (thereby conflating Hardy's distinction between the beautiful and the aesthetic). He thus proposed two definitive criteria: the principle of minimal completeness and the principle of maximal applicability. King illustrated both principles using the Pythagorean proof of the irrationality of $\sqrt{2}$. The principle of minimal completeness, in effect, functions as a super-class to Hardy's aesthetic qualities. However, King's principle of maximal applicability resonates more with Hardy's notions of significance, depth and generality.

Finally, David Wells's (1990) survey of contemporary mathematicians has most convincingly illuminated the subjectivity question. He asked the readers of *The Mathematical Intelligencer* to rate, on a scale of one to ten, twenty-four theorems according to their mathematical beauty. From the seventy-six responses, many from top mathematicians mostly from North America, he drew a number of inferences. First, mathematicians do not always agree on their aesthetic judgements – at least not in terms of evaluating the beauty of theorems.

Wells identified many factors that contribute to the differences in judgement: field of interest; preferences for certain mathematical entities such as problems, proofs or theorems; past experiences or associations with particular theorems; even mood. He also pointed out that aesthetic judgements change over time: this was particularly evident in the rating of Euler's formula, which was historically considered "the most beautiful formula of mathematics" (p. 38), but is now, according to Wells's respondents at least, considered too obvious even to elicit an aesthetic response.

The inferences made by Wells correspond to a contextualist view of aesthetic appreciation and are summed up by this respondent: "beauty, even in mathematics, depends upon historical and cultural contexts, and therefore tends to elude numerical interpretation" (p. 39). Indeed, John von Neumann had already spoken of the phenomenon of mathematical 'styles' back in 1947, arguing that, it is "hardly possible to believe in the existence of an absolute, immutable concept of mathematical rigor, dissociated from all human experience" (p. 190). He used as evidence the changes in styles of mathematical proofs and fashionable areas of interest over the past two millennia.

One might wonder why these changes in style appear so much less dramatic than the ones found in the arts. Are the styles necessarily more confined in mathematics, owing to the handful of aesthetic commitments that ultimately define the discipline? Or does the study of mathematics attract a small enough number of like-minded people that aesthetic revolutionaries such as Picasso, Pollock or Cage do not have mathematical equivalents?

Some Final Comments

It is certainly tempting to wonder why the twentieth century witnessed such an explosion in thinking about the mathematical aesthetic, if only to help predict what might be in store for the twenty-first. Will this book be unawarely documenting the close of an active period of investigation or will it serve as a springboard for further, fruitful inquiry?

It would be hard to overlook the fact that at turn of the twentieth century, not long after the discovery of non-Euclidean geometries and just as Cantor's work on trigonometric series and the continuum was emerging, foundational concerns were mounting and questions about axiomatisation were beginning to press. These concerns would incite mathematicians to begin seriously inspecting the nature of mathematics and for some, such as David Hilbert, Herman Weyl and L. E. J. Brouwer, to turn their attention to 'meta-mathematical' questions, albeit with markedly different responses.

Few mathematicians since the ancient Greeks have stepped back from the exhilarating momentum of creating mathematics to consider larger epistemological questions. Hilbert's famous list of unsolved problems, which essentially came to define much of what would be considered 'interesting' to work on, also came at the turn of the twentieth century. His lengthy list either prompted or nourished a broader consideration of the whole field of mathematics – its goals, methods, and successes – yielding yet more 'meta'-mathematical thinking, which could hardly ignore the important aesthetic dimension of mathematics.

It is striking to us that mathematicians often mention 'beauty', yet there seems to be a relative dearth of further amplification. One might have expected those past mathematicians who thought in these terms to have been capable of developing ideas of, say harmony, proportion, fit, rhythm, etc, more precisely. To some extent, this is what Hardy (1940) tried to do, though only by connecting 'beauty' to other barely less opaque terms such as 'elegance', 'depth', 'seriousness' and 'significance'. We do see instances here and there, for instance with Alfred North Whitehead (on rhythm), with Warwick Sawyer (on pattern) and in an overly-mathematised attempt by George Birkhoff.

But it may well be that many mathematicians simply do not consider this to be a serious enterprise, one worthy of their time and attention. Even

Hardy (1940) expressed a sense that such reflection 'about' mathematics (offering a different sense of 'meta'-mathematical activity) is not really the preferred activity or even the very business of mathematicians.

> It is a melancholy experience for a professional mathematician to find himself writing about mathematics. The function of a mathematician is to do something, to prove new theorems, to add to mathematics, and not to talk about what he or other mathematicians have done. (p. 61)

There seem to be some 'inevitable' combinations of aesthetic words that are mathematically invoked as if conjoined: for instance, beauty *and* elegance, perfection *and* beauty. 'Elegance', in particular, seems to have been co-opted by mathematicians in their rather restricted aesthetic language, as conveying a sense of both succinctness and sophistication. In ordinary parlance, 'elegance' might be seen as a classical, class-ridden term – not so much socio-economic 'class' perhaps as intellectual 'class' (though Bertrand Russell, for example, certainly partook of both). Of course, there must be a sociological proviso here – it was only very few (privileged) Greeks, and then for a long time a very few (privileged) other individuals, who could sit and think as opposed to practice or teach.

There might also be more subtle reasons for this explosion of aesthetic consideration and writing. It was also around the turn of the last century that the field of mathematics made its final separation from the sciences, its increasing abstractions having less and less to do with the kind of questions that drove the development of calculus, for example. Mathematics was carving itself out as a distinct, self-sufficient field with famously little to do with the 'real' world. But how, then, could it justify its existence? This was a question that those both inside and outside the tall, opaque walls guarding the mathematical terrain asked.

Perhaps this question prompted some mathematicians to search for some varied connections – ones that many ancient Greek mathematicians would have assumed – between mathematics and the arts, another field which offers few practical applications, though is admired on the whole for its display of creativity and its production of aesthetically pleasing artifacts. Like the modernist art movement, which was burgeoning with a sense of art for art's sake *(l'art pour l'art)* during this time in the early twentieth century, mathematics was now being done for mathematics' sake.

One instance of this modernist mathematical ethos is apparent in van der Waerden's (1930/1991) *Modern Algebra,* in which any questions of algebra's utility had completely vanished (see Chapter 8). John Dee's (1570) survey of various branches of 'the tree of mathematics' in his *Mathematicall præface* referred positively to his identification of which parts of mathematics were, to use his term, 'commodious': van der Waerden felt no such compunction.

This twentieth-century expansion of interest in the aesthetic did not only occur in mathematics, of course. As Denis Donoghue (2003) observes:

> Interest in beauty and aesthetics was greatly stimulated in the first
> years of the twentieth century — as C. K. Ogden, I. A. Richards,
> and James Wood noted in *The Foundations of Aesthetics* (1922) —
> by a wider knowledge of non-European art, especially of Eastern
> and primitive art, and by the rapid development of psychology as
> an accredited practice. (p. 36)

Additionally, in the latter part of the twentieth century, developments mostly outside mathematics itself may have further contributed to this most recent explosion within mathematics. For example, scholars in a number of different fields have become increasingly attracted to sociobiologist E. O. Wilson's (1998) notion of 'consilience'. In his view, unification or connection is sought among the increasingly-fractured disciplines and ways of knowing. He is also concerned, more generally, with the breaking down of long-standing dichotomies between *mind* and *body,* between *rational* and *emotional,* between *logical* and *intuitive.* Although few scholars have contributed to the search for affinities in the still often-isolated and inhospitable world of mathematics, these intellectual changes are at least supportive of such endeavours, as we hope this book will show.

Notes

[1] Writing about eighty years after this event, Oxford University historian Anthony à Wood observed:

> Many MSS, guilty of no other superstition than red letters in their
> fronts or title, were either condemned to the fire or jakes. [...] sure
> I am that such books wherein appeared Angles, or Mathematical
> Diagrams, were thought sufficient to be destroyed, because
> accounted Popish, or diabolical, or both.
>
>   What was done to the public Library I shall elsewhere shew: as
> for those belonging to Colleges, they suffered the same fate almost
> as the public, though not in so gross a manner. From Merton Coll.
> Library a cart load of MSS and above were taken away, such that
> contained the Lucubrations (chiefly of controversial Divinity,
> Astronomy and Mathematicks) of divers of the learned Fellows
> thereof, in which Studies they in the two last centuries obtained
> great renown. (in Gutch, 1796, pp. 106-107)

[2] This twentieth-century return to the question of the psychology of the mathematician connects for us with thirteenth-century Henry of Ghent's potent phrase 'the melancholy disposition of the mathematical mind'. It was coined in the light of much Aristotelean writing and the then-dominant Galenic theory of 'humours', a means of linking human psychology with the cosmos. It also relates closely to Albrecht Dürer's famous engraving *Melencolia I* – see Yates (1979). This is a topic we explore further in the closing chapter of this book.

[3] The Pythagoreans, who celebrated the Muses as "the keepers of the knowledge of harmony and the principles of the universe which allowed access to the ever-lasting gods" (see Comte, 1994, p. 135), would have been delighted by the trust that scientists, and mathematicians, have come to place on this aesthetic muse.

Section A

The Mathematician's Art

Introduction to Section A

The three chapters of Section A, *The Mathematician's Art,* all written by research mathematicians, provide satisfying glimpses into present-day aesthetic dimensions in mathematical work. By focusing on the methods, processes, experiences and goals of their own mathematical endeavours, in fields such as geometry, number theory, real and complex analysis, combinatorics and topology, these mathematicians offer a broad view of modern manifestations of the ancient affinity of our book title. In particular, they ground the general theory-driven accounts of Poincaré and Hadamard (discussed in Chapter α) in their varied, day-to-day practices of working in mathematics.

Jonathan Borwein, in Chapter 1, draws on his own research in number theory and analysis to show how aesthetic notions permeate both pure and applied mathematics. His examples illustrate aesthetic imperatives interacting with utility, with intuition and with the way they shape his own mathematical experiences. His examples, drawing as they do on diverse, contemporary topics as well as 'hot' methods (i.e. involving computer technology), provide a welcome and up-to-date perspective on a topic where the same 'classical' examples are frequently cited. Borwein is also interested in tracing out how aesthetic criteria change over time and how these changes manifest themselves in the concerns and discoveries of mathematicians.

In Chapter 2, Doris Schattschneider provides a discussion of her candidates for elegant statements, beautiful proofs and some important paradigms of mathematical technique, such as Fubini's principle or Dirichlet's principle. These paradigms serve as attractors, condensed and powerful modes of arguing which signal their generative potential for the future, as well as attest to their value in the past. Using examples taken primarily from plane geometry and combinatorics, Schattschneider identifies various characteristics of mathematical proofs that can provoke aesthetic pleasure. She devotes the last section of her chapter to the aesthetic of *doing* mathematics, examining the motivational dimension of that aesthetic in mathematical activity.

David Henderson and Daina Taimina, in Chapter 3, take up Schattschneider's focus on the *doing* of mathematics, with a particular emphasis on the aesthetics of generating and experiencing mathematical understanding. Using examples from hyperbolic geometry, topology and also nineteenth-century mechanical devices with sophisticated mathematical features, Henderson and Taimina illustrate the way in which their own understanding of meaning in mathematics emerges from attempts to connect fundamental intuitions to subtle mathematical ideas and claims. The aesthetic experiences they describe all involve sense-based encounters with mathematical ideas and, in sharing them, offer compelling and sophisticated images, objects and models that they have either created or derived.

CHAPTER 1
Aesthetics for the Working Mathematician

Jonathan M. Borwein

> If my teachers had begun by telling me that mathematics was pure play with presuppositions, and wholly in the air, I might have become a good mathematician, because I am happy enough in the realm of essence. But they were over-worked drudges, and I was largely inattentive, and inclined lazily to attribute to incapacity in myself or to a literary temperament that dullness which perhaps was due simply to lack of initiation. (Santayana, 1944, p. 238)

Most research mathematicians neither think deeply about nor are terribly concerned with either pedagogy or the philosophy of mathematics. Nonetheless, as I hope to indicate, aesthetic notions have always permeated (pure and applied) mathematics. And the top researchers have always been driven by an aesthetic imperative. Many mathematicians over time have talked about the 'elegance' of certain proofs or the 'beauty' of certain theorems, but my analysis goes deeper: I aim to show how the aesthetic imperative interacts with utility and intuition, as well as indicate how it serves to shape my own mathematical experiences. These analyses, rather than being retrospective and passive, will provide a living account of the aesthetic dimension of mathematical work.

> We all believe that mathematics is an art. The author of a book, the lecturer in a classroom tries to convey the structural beauty of mathematics to his readers, to his listeners. In this attempt, he must always fail. Mathematics is logical to be sure, each conclusion is drawn from previously derived statements. Yet the whole of it, the real piece of art, is not linear; worse than that, its perception should be instantaneous. We all have experienced on some rare occasions the feeling of elation in realizing that we have enabled our listeners to see at a moment's glance the whole architecture and all its ramifications. (Emil Artin, in Murty, 1988, p. 60)

I shall similarly argue for aesthetics before utility. Through a suite of examples drawn from my own research and interests, I aim to illustrate how and what this means on the front lines of research. I also will argue that the opportunities to evoke the mathematical aesthetic in research and teaching are almost boundless – at all levels of the curriculum. (An excellent middle-school illustration, for instance, is described in Sinclair, 2001.)

In part, this is due to the increasing power and sophistication of visualisation, geometry, algebra and other mathematical software. Indeed, by drawing on 'hot topics' as well as 'hot methods' (i.e. computer technology),

I also provide a contemporary perspective which I hope will complement the more classical contributions to our understanding of the mathematical aesthetic offered by writers such as G. H. Hardy and Henri Poincaré (as discussed in Chapter α).

Webster's dictionary (1993, p. 19) first provides six different meanings of the word 'aesthetic', used as an adjective. However, I want to react to these two definitions of 'aesthetics', used as a noun:

1. The branch of philosophy dealing with such notions as the beautiful, the ugly, the sublime, the comic, etc., as applicable to the fine arts, with a view to establishing the meaning and validity of critical judgments concerning works of art, and the principles underlying or justifying such judgments.
2. The study of the mind and emotions in relation to the sense of beauty.

Personally, for my own definition of the aesthetic, I would require (unexpected) simplicity or organisation in apparent complexity or chaos – consistent with views of Dewey (1934), Santayana (1944) and others. I believe we need to integrate this aesthetic into mathematics education at every level, so as to capture minds for other than utilitarian reasons. I also believe detachment to be an important component of the aesthetic experience: thus, it is important to provide some curtains, stages, scaffolding and picture frames – or at least their mathematical equivalents. Fear of mathematics certainly does not hasten an aesthetic response.

Gauss, Hadamard and Hardy

Three of my personal mathematical heroes, very different individuals from different times, all testify interestingly on the aesthetic and the nature of mathematics.

Gauss

Carl Friedrich Gauss is claimed to have once confessed, "I have had my results for a long time, but I do not yet know how I am to arrive at them" (in Arber, 1954, p. 47). [1] One of Gauss's greatest discoveries, in 1799, was the relationship between the lemniscate sine function and the arithmetic–geometric mean iteration. This was based on a purely computational observation. The young Gauss wrote in his diary that "a whole new field of analysis will certainly be opened up" (*Werke*, X, p. 542; in Gray, 1984, p. 121).

He was right, as it pried open the whole vista of nineteenth-century elliptic and modular function theory. Gauss's specific discovery, based on tables of integrals provided by Scotsman James Stirling, was that the reciprocal of the integral

$$\frac{2}{\pi} \int_0^1 \frac{1}{\sqrt{1 - t^4}} \, dt$$

agreed numerically with the limit of the rapidly convergent iteration given by setting $a_0 := 1$, $b_0 := \sqrt{2}$ and then computing:

$$a_{n+1} := \frac{a_n + b_n}{2}, \qquad b_{n+1} := \sqrt{a_n b_n}.$$

It transpires that the two sequences $\{a_n\}$, $\{b_n\}$ have a common limit of 1.1981402347355922074...

Which object, the integral or the iteration, is the more familiar and which is the more elegant – then and now? Aesthetic criteria change with time (and within different cultures) and these changes manifest themselves in the concerns and discoveries of mathematicians. Gauss's discovery of the relationship between the lemniscate function and the arithmetic–geometric mean iteration illustrates how the traditionally preferred 'closed form' (here, the integral form) of equations have yielded centre stage, in terms both of elegance and utility, to recursion. This parallels the way in which biological and computational metaphors (even 'biology envy') have now replaced Newtonian mental images, as described and discussed by Richard Dawkins (1986) in his book *The Blind Watchmaker*.

In fact, I believe that mathematical thought patterns also change with time and that these in turn affect aesthetic criteria – not only in terms of what counts as an interesting problem, but also what methods the mathematician can use to approach these problems, as well as how a mathematician judges their solutions. As mathematics becomes more 'biological', and more computational, aesthetic criteria will continue to change.

Hadamard

A constructivist, experimental and aesthetically-driven rationale for mathematics could hardly do better than to start with Hadamard's claim that:

> The object of mathematical rigor is to sanction and legitimize the conquests of intuition, and there was never any other object for it.
> (in Pólya, 1981, p. 127)

Jacques Hadamard was perhaps the greatest mathematician other than Poincaré to think deeply and seriously about cognition in mathematics. He is quoted as saying, "in arithmetic, until the seventh grade, I was last or nearly last" (in MacHale, 1993, p. 142). Hadamard was co-prover (independently with Charles de la Vallée Poussin, in 1896) of the Prime Number theorem (the number of primes not exceeding n is asymptotic to $n/\log n$), one of the culminating results of nineteenth-century mathematics and one that relied on much preliminary computation and experimentation. He was also the author of *The Psychology of Invention in the Mathematical Field* (1945), a book that still rewards close inspection.

Hardy's *Apology*

Correspondingly, G. H. Hardy, the leading British analyst of the first half of the twentieth century, was also a stylish author who wrote compellingly in defence of pure mathematics. He observed that:

> All physicists and a good many quite respectable mathematicians are contemptuous about proof. (1945/1999, pp. 15-16)

His memoir, entitled *A Mathematician's Apology,* provided a spirited defence of beauty over utility:

> Beauty is the first test. There is no permanent place in the world for ugly mathematics. (1940, p. 84)

That said, although the sentiment behind it being perfectly understandable from an anti-war mathematician in war-threatened Britain, Hardy's claim that real mathematics is almost wholly useless has been over-played and, to my mind, is now very dated, given the importance of cryptography and other pieces of algebra and number theory devolving from very pure study.

In his tribute to Srinivasa Ramanujan entitled *Ramanujan: Twelve Lectures on Subjects Suggested by His Life and Work,* Hardy (1945/1999) offered the so-called 'Skewes number' as a "striking example of a false conjecture" (p. 15). The logarithmic integral function, written Li(x), is specified by:

$$Li(x) = \int_0^x \frac{1}{\log(t)} \, dt$$

Li(x) provides a very good approximation to the number of primes that do not exceed x. For example, Li(10^8) = 5,762,209.375…, while the number of primes not exceeding 10^8 is 5,761,455. It was conjectured that the inequality

> Li(x) > the number of primes not exceeding x

holds for all x and, indeed, it does so for many x. In 1933, Skewes showed the first explicit crossing occurs before $10^{10^{10^{34}}}$. This has been reduced to a relatively tiny number, a mere 10^{1167} (and, most recently, even lower), though one still vastly beyond direct computational reach.

Such examples show forcibly the limits on numerical experimentation, at least of a naïve variety. Many readers will be familiar with the 'law of large numbers' in statistics. Here, we see an instance of what some number theorists (e.g. Guy, 1988) call the 'strong law of small numbers': *all small numbers are special,* many are primes and direct experience is a poor guide. And sadly (or happily, depending on one's attitude), even 10^{1167} may be a small number.

Research Motivations and Goals

As a computational and experimental pure mathematician, my main goal is *insight.* Insight demands speed and, increasingly, parallelism (see Borwein and Borwein, 2001, on the challenges for mathematical computing). The mathematician's 'aesthetic buzz' comes not only from simply contemplating a beautiful piece of mathematics, but, additionally, from achieving insight. The computer, with its capacities for visualisation and computation, can encourage the aesthetic buzz of insight, by offering the mathematician the possibility of visual contact with mathematics and by allowing the mathematician to experiment with, and thus to become intimate with, mathematical ideas, equations and objects.

What is 'easy' is changing and I see an exciting merging of disciplines, levels and collaborators. Mathematicians are more and more able to:

- marry theory and practice, history and philosophy, proofs and experiments;
- match elegance and balance with utility and economy;
- inform all mathematical modalities computationally – analytic, algebraic, geometric and topological.

This is leading us towards what I term an *experimental methodology* as a philosophy and a practice (Borwein and Corless, 1999). This methodology is based on the following three approaches:

- meshing computation and mathematics, so that intuition is acquired;
- visualisation – three is a lot of dimensions and, nowadays, we can exploit pictures, sounds and haptic stimuli to get a 'feel' for relationships and structures (see also Chapter 7);
- 'exception barring' and 'monster barring' (using the terms of Lakatos, 1976).

Two particularly useful components of this third approach include graphical and randomised checks. For example, comparing $2\sqrt{y} - y$ and $-\sqrt{y}\ln(y)$ (for $0 < y < 1$) pictorially is a much more rapid way to divine which is larger than by using traditional analytic methods. Similarly, randomised checks of equations, inequalities, factorisations or primality can provide enormously secure knowledge or counter-examples when deterministic methods are doomed. As with traditional mathematical methodologies, insight and certainty are still highly valued, yet achieved in different ways.

Pictures and symbols

> If I can give an abstract proof of something, I'm reasonably happy. But if I can get a concrete, computational proof and actually produce numbers I'm much happier. I'm rather an addict of doing things on the computer, because that gives you an explicit criterion of what's going on. I have a visual way of thinking, and I'm happy if I can see a picture of what I'm working with. (John Milnor, in Regis, 1986, p. 78)

I have personally had this experience, in the context of studying the distribution of zeroes of the Riemann zeta function. Consider more explicitly the following image (see Figure 1), which shows the densities of zeroes for polynomials in powers of x with -1 and 1 as coefficients (they are manipulable at: www.cecm.sfu.ca/interfaces/). All roots of polynomials, up to a given degree, with coefficients of either -1 or 1 have been calculated by permuting through all possible combinations of polynomials, then solving for the roots of each. These roots are then plotted on the complex plane (around the origin).

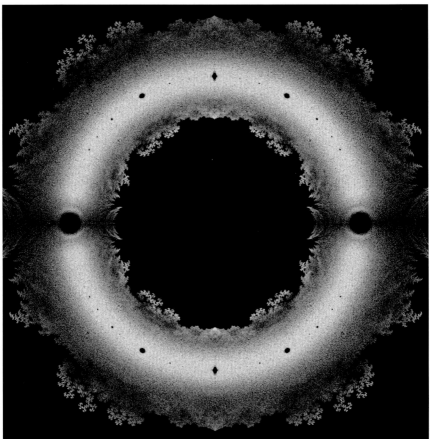

Figure 1: Density of zeroes for polynomials with coefficients of -1 and 1

In this case, graphical output from a computer allows a level of insight no amount of numbers could.

Some colleagues and I have been building educational software with these precepts embedded, such as *LetsDoMath* (see: www.mathresources. com). The intent is to challenge students honestly (e.g. through allowing subtle explorations within John Conway's 'Game of Life'), while making things tangible (e.g. 'Platonic solids' offers virtual manipulables that are more robust and expressive than the standard classroom solids).

Evidently, though, symbols are often more reliable than pictures. The picture opposite purports to give evidence that a solid can fail to be polyhedral at only one point. It shows the steps up to pixel level of inscribing a

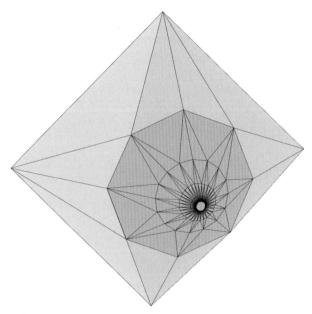

Figure 2: A misleading picture

regular 2^{n+1}-gon at height 2^{1-n}. However, ultimately, such a construction fails and produces a right circular cone. The false evidence in this picture held back a research project for several days – and might have derailed it.

Two Things about √2 and One Thing about π

Remarkably, one can still find new insights in the oldest areas. I discuss three examples of this. The first involves a new proof of the irrationality of √2 and the way in which it provides insight into a previously known result. The second invokes the strange interplay between rational and irrational numbers. Finally, the third instance reveals how the computer can make opaque some properties that were previously transparent, and *vice versa.*

Irrationality

Below is a graphical representation of Tom Apostol's (2000) lovely new geometric proof of the irrationality of √2. This example may seem routine at first, with respect to the literature on the mathematical aesthetic. Writers such as Hardy (1940), King (1992) and Wells (1990) have also talked about the beauty of quadratics such as √2. These writers have emphasised aesthetic criteria (such as economy and unexpectedness) that contribute to that judgement of beauty. On the other hand, Apostol's new proof, prefigured in others, shows how aesthetics can also serve to *motivate* mathematical inquiry.

PROOF Consider the *smallest* right-angled isosceles triangle with integer sides. Circumscribe a circle of length equal to the vertical side and construct the tangent to the circle where the hypotenuse cuts it (see Figure 3). The *smaller* isosceles triangle once again has integer sides.

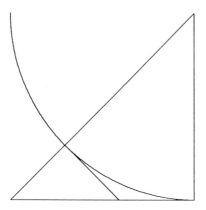

Figure 3: The square root of two is irrational

The proof is lovely because it offers new insight into a result that was first proven over two thousand years ago. It also verges on being a 'proof without words' (Nelsen, 1993), proofs which are much admired – yet infrequently encountered and not always trusted – by mathematicians (see Brown, 1999). Apostol's work demonstrates how mathematicians are not only motivated to find ground-breaking results, but that they also strive for better ways to say things or to show things, as Gauss was surely doing when he worked out his fourth, fifth and sixth proof of the law of quadratic reciprocity.

Rationality

By a variety of means, including the one above, we know that the square root of two is irrational. But mathematics is always full of surprises: $\sqrt{2}$ can also *make* things rational (a case of two wrongs making a right?).

$$\left(\sqrt{2}^{\sqrt{2}}\right)^{\sqrt{2}} = \sqrt{2}^{(\sqrt{2}\sqrt{2})} = \sqrt{2}^2 = 2.$$

Hence, by the principle of the excluded middle:

$$\text{Either} \quad \sqrt{2}^{\sqrt{2}} \in Q \quad \text{or} \quad \sqrt{2}^{\sqrt{2}} \notin Q.$$

In either case, we can deduce that there are irrational numbers a and b with a^b rational. But how do we know which ones? One may build a whole mathematical philosophy project around this. Yet, as *Maple* (the computer algebra system) confirms:

setting $\alpha := \sqrt{2}$ and $\beta := 2\ln_2 3$ yields $\alpha^\beta = 3$.

This illustrates nicely that verification is often easier than discovery. (Similarly, the fact that multiplication is easier than factorisation is at the base of secure encryption schemes for e-commerce.)

π and two integrals

Even *Maple* knows $\pi \neq 22/7$, since:

$$0 < \int_0^1 \frac{(1-x)^4 \, x^4}{1+x^2} \, dx = \frac{22}{7} - \pi.$$

Nevertheless, it would be prudent to ask 'why' *Maple* is able to perform the evaluation and whether to trust it. In contrast, *Maple* struggles with the following *sophomore's dream:*

$$\int_0^1 \frac{1}{x^x} \, dx = \sum_{n=1}^{\infty} \frac{1}{n^n}.$$

Students asked to confirm this typically mistake numerical validation for symbolic proof.

Again, we see that computing adds reality, making the abstract concrete, and makes some hard things simple. This is strikingly the case with Pascal's Triangle. Figure 4 (from: www.cecm.sfu.ca/interfaces/) affords an emphatic example where deep fractal structure is exhibited in the elementary binomial coefficients.

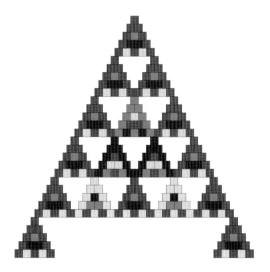

Figure 4: Thirty rows of Pascal's triangle (modulo five)

Berlinski (1997) comments on some of the effects of such visual–experimental possibilities in mathematics:

> The computer has in turn changed the very nature of mathematical experience, suggesting for the first time that mathematics, like physics, may yet become an empirical discipline, a place where things are discovered because they are seen. (p. 39)

Berlinski (1995) had earlier suggested, in his book *A Tour of the Calculus,* that there will be long-term effects:

> The body of mathematics to which the calculus gives rise embodies a certain swashbuckling style of thinking, at once bold and dramatic, given over to large intellectual gestures and indifferent, in large measure, to any very detailed description of the world. It is a style that has shaped the physical but not the biological sciences, and its success in Newtonian mechanics, general relativity, and quantum mechanics is among the miracles of mankind. But the era in thought that the calculus made possible is coming to an end. Everyone feels this is so, and everyone is right. (p. xiii)

π and Its Friends

My research on π with my brother, Peter Borwein, also offers aesthetic and empirical opportunities. In this example, my personal fascinations provide compelling illustrations of an aesthetic imperative in my own work. I first discuss the algorithms I have co-developed to compute the digits of π. These algorithms, which consist of simple algebraic equations, have made it possible for researchers to compute its first 3×2^{36} digits. I also discuss some of the methods and algorithms I have used to gain insight into relationships involving π.

A quartic algorithm (Borwein and Borwein, 1984)

The next algorithm I present grew out of work of Ramanujan. Set $a_0 = 6 - 4\sqrt{2}$ and $y_0 = \sqrt{2} - 1$. Iterate:

$$(1) \qquad y_{k+1} = \frac{1 - (1 - y_k^4)^{1/4}}{1 + (1 - y_k^4)^{1/4}}$$

$$(2) \qquad a_{k+1} = a_k(1 + y_{k+1})^4 - 2^{2k+3}\, y_{k+1}(1 + y_{k+1} + y_{k+1}^2)$$

Then the sequence $\{a_k\}$ converges *quartically* to $1/\pi$.

There are nineteen pairs of simple algebraic equations (1, 2) as k ranges from 0 to 18. After seventeen years, this still gives me an aesthetic buzz. Why? With less than one page of equations, I have a tool for computing a number that differs from π (the most celebrated transcendental number) only after seven hundred billion digits. It is not only the economy of the tool that delights me, but also the stirring idea of 'almost-ness' – that even after seven hundred billion digits we still cannot nail π. The difference might seem trivial, but mathematicians know that it is not and they continue to improve their algorithms and computational tools.

This iteration has been used since 1986, with the Salamin–Brent scheme, by David Bailey (at the Lawrence Berkeley Labs) and by Yasumasa Kanada (in Tokyo). In 1997, Kanada computed over 51 billion digits on a Hitachi supercomputer (18 iterations, 25 hrs on 210 cpus). His penultimate world record was 2^{36} digits in April, 1999. A billion (2^{30}) digit computation has been performed on a single Pentium II PC in less than nine days. The present record is 1.24 trillion digits, computed by Kanada in December 2002 using quite different methods, and is described in my new book, co-authored with David Bailey (2003).

The fifty-billionth decimal digit of π *or* of $1/\pi$ is 04<u>2</u>! And after eighteen billion digits, the string 0123456789 has finally appeared and so Brouwer's famous intuitionist example *now* converges. [2] (Details such as this about π can be found at: www.cecm.sfu.ca/personal/jborwein/pi_cover.html.) From a probability perspective, such questions may seem uninteresting, but they continue to motivate and amaze mathematicians.

A further taste of Ramanujan

G. N. Watson, in discussing his response to similar formulae of the wonderful Indian mathematical genius Srinivasa Ramanujan, describes:

> a thrill which is indistinguishable from the thrill which I feel when I enter the Sagrestia Nuova of the Capelle Medicee and see before me the austere beauty of [the four statues representing] 'Day,' 'Night,' 'Evening,' and 'Dawn' which Michelangelo has set over the tombs of Giuliano de' Medici and Lorenzo de' Medici. (in Chandrasekhar, 1987, p. 61)

One of these is Ramanujan's remarkable formula, based on the elliptic and modular function theory initiated by Gauss.

$$\frac{1}{\pi} = \frac{2\sqrt{2}}{9801} \sum_{k=0}^{\infty} \frac{(4k)!(1103 + 26390k)}{(k!)^4 \; 396^{4k}} .$$

Each term of this series produces an additional $\overline{eight}$ correct digits in the result – and only the ultimate multiplication by $\sqrt{2}$ is not a *rational* operation. Bill Gosper used this formula to compute seventeen million terms of the continued fraction for π in 1985. This is of interest, because we still cannot prove that the continued fraction for π is unbounded. Again, every one *knows* that this is true.

That said, Ramanujan preferred related explicit forms for approximating π, such as the following:

$$\frac{\log(640320^3)}{\sqrt{163}} = 3.1415926535897930\underline{0164} \approx \pi.$$

This equation is correct until the underlined places. *Inter alia,* the number e^{π} is the easiest transcendental to fast compute (by elliptic methods). One 'differentiates' $e^{-\pi t}$ to obtain algorithms such as the one above for π, via the arithmetic–geometric mean.

Integer relation detection

I make a brief digression to describe what integer relation detection methods do. (These may be tried at: www.cecm.sfu.ca/projects/IntegerRelations/.) I then apply them to π (see Borwein and Lisonek, 2000).

> DEFINITION A vector $(x_1, x_2, \dots, x_n)$ of real numbers *possesses an integer relation,* if there exist integers a_i (not all zero) with:
>
> $$a_1 x_1 + a_2 x_2 + \dots + a_n x_n = 0$$
>
> PROBLEM Find a_i if such integers exist. If not, obtain lower 'exclusion' bounds on the size of possible a_i.
>
> SOLUTION For $n = 2$, *Euclid's algorithm* gives a solution. For $n \geq 3$, Euler, Jacobi, Poincaré, Minkowski, Perron and many others sought methods. The *first general algorithm* was found (in 1977) by Ferguson and Forcade. Since 1977, one has many variants: I will mainly be talking about two algorithms, LLL ('Lenstra, Lenstra and Lovász'; also available in *Maple* and *Mathematica*) and PSLQ ('Partial sums using matrix LQ decomposition', 1991; *parallelised,* 1999).

Integer relation detection was recently ranked among:

> the 10 algorithms with the greatest influence on the development and practice of science and engineering in the 20th century. (Dongarra and Sullivan, 2000, p. 22)

It could be interesting for the reader to compare these algorithms with the theorems on the list of the most 'beautiful' theorems picked out by Wells (1990) in his survey, in terms of criteria such as applicability, unexpectedness and fruitfulness.

Determining whether or not a number is algebraic is one problem that can be attacked using integer relation detection. Asking about algebraicity is handled by computing α to sufficiently high precision ($O(n = N^2)$) and applying LLL or PSLQ to the vector $(1, \alpha, \alpha^2, \dots, \alpha^{N-1})$. Solution integers a_i are coefficients of a polynomial likely satisfied by α. If one has computed α to $n + m$ digits and run LLL using n of them, one has m digits to confirm the result heuristically. I have never seen this method return an honest 'false positive' for $m > 20$, say. If no relation is found, exclusion bounds are obtained, saying, for example, that any polynomial of degree less than N must have the Euclidean norm of its coefficients in excess of L (often astronomical). If we know or suspect an identity exists, then integer relations methods are very powerful. Let me illustrate this in the context of approximating π.

Machin's formula

We use *Maple* to look for the linear dependence of the following quantities:

[arctan(1), arctan(1/5), arctan(1/239)]

and 'recover' [1, −4, 1]. In other words, we can establish the following equation:

$$\pi/4 = 4\arctan(1/5) - \arctan(1/239).$$

Machin's formula was used on all serious computations of π from 1706 (a hundred digits) to 1973 (a million digits), as well as more abstruse but similar formulae used in creating Kanada's present record. After 1980, the methods described above started to be used instead.

Dase's formula

Again, we use *Maple* to look for the linear dependence of the following quantities:

[$\pi/4$, arctan(1/2), arctan(1/5), arctan(1/8)].

and recover [−1, 1, 1, 1]. In other words, we can establish the following equation:

$$\pi/4 = \arctan(1/2) + \arctan(1/5) + \arctan(1/8).$$

This equation was used by Dase to compute two hundred digits of π in his head in perhaps the greatest feat of mental arithmetic ever – 1/8 is apparently better than 1/239 (as in Machin's formula) for this purpose.

Who was Dase? Another burgeoning component of modern research and teaching life is having access to excellent data bases, such as the MacTutor History Archive maintained at: www-history.mcs.st-andrews.ac.uk (alas, not all sites are anywhere near so accurate and informative as this one). One may find details there on almost all of the mathematicians appearing in this chapter. I briefly illustrate its value by showing verbatim what it says about Dase.

> Zacharias Dase (1824–1861) had incredible calculating skills but little mathematical ability. He gave exhibitions of his calculating powers in Germany, Austria and England. While in Vienna in 1840 he was urged to use his powers for scientific purposes and he discussed projects with Gauss and others.

> Dase used his calculating ability to calculate π to 200 places in 1844. This was published in Crelle's Journal for 1844. Dase also constructed 7 figure log tables and produced a table of factors of all numbers between 7 000 000 and 10 000 000.

Gauss requested that the Hamburg Academy of Sciences allow Dase to devote himself full-time to his mathematical work but, although they agreed to this, Dase died before he was able to do much more work.

Pentium farming

I finish this sub-section with another result obtained through integer relations methods or, as I like to call it, 'Pentium farming'. Bailey, Borwein and Plouffe (1997) discovered a series for π (and corresponding ones for some other *polylogarithmic* constants), which somewhat disconcertingly allows one to compute hexadecimal digits of π *without* computing prior digits. (This feels like magic, being able to tell the seventeen-millionth digit of π, say, without having to calculate the ones before it; it is like seeing God reach her hand deep into π.)

The algorithm needs very little memory and no multiple precision. The running time grows only slightly faster than linearly in the order of the digit being computed. The key, found by PSLQ as described above, is:

$$\pi = \sum_{k=0}^{\infty} \left(\frac{1}{16}\right)^{k} \left(\frac{4}{8k+1} - \frac{2}{8k+4} - \frac{1}{8k+5} - \frac{1}{8k+6}\right).$$

Knowing an algorithm would follow, Bailey, Borwein and Plouffe spent several months hunting by computer for such a formula. Once found, it is easy to prove in *Mathematica,* in *Maple* or by hand – and provides a very nice calculus exercise.

This was a most successful case of *reverse mathematical engineering* and is entirely practicable. In September 1997, Fabrice Bellard (at INRIA) used a variant of this formula to compute one hundred and fifty-two binary digits of π, starting at the *trillionth* (10^{12}) place. This took twelve days on twenty work-stations working in parallel over the internet. In August 1998, Colin Percival (Simon Fraser University, age 17) finished a 'massively parallel' computation of the *five-trillionth bit* (using twenty-five machines at roughly ten times the speed of Bellard). In *hexadecimal notation,* he obtained:

07E45 733CC790B5B5979.

The corresponding binary digits of π starting at the forty-trillionth bit are:

0 0000 1111 1001 1111.

By September 2000, the quadrillionth bit had been found to be the digit 0 (using 250 cpu years on a total of one thousand, seven hundred and thirty-four machines from fifty-six countries). Starting at the 999,999,999,999,997th bit of π, we find:

11100 0110 0010 0001 0110 1011 0000 0110.

Solid and Discrete Geometry – and Number Theory

Although my own primary research interests are in numerical, classical and functional analysis, I find that the fields of solid and discrete geometry, as well as number theory, offer many examples of the kinds of concrete, insightful ideas I value. In the first example, I argue for the computational affordances available to study of solid geometry. I then discuss the genesis of an elegant proof in discrete geometry. Finally, I illustrate a couple of deep results in partition theory.

de Morgan

Augustus de Morgan, one of the most influential educators of his period and first president of the London Mathematical Society, wrote:

> Considerable obstacles generally present themselves to the beginner, in studying the elements of Solid Geometry, from the practice which has hitherto uniformly prevailed in this country, of never submitting to the eye of the student, the figures on whose properties he is reasoning, but of drawing perspective representations of them upon a plane. [...] I hope that I shall never be obliged to have recourse to a perspective drawing of any figure whose parts are not in the same plane. (in Rice, 1999, p. 540)

His comment illustrates the importance of concrete experiences with mathematical objects, even when the ultimate purpose is to abstract. There is a sense in which insight lies in physical manipulation. I imagine that de Morgan would have been happier using JavaViewLib (see: www.cecm.sfu. ca/interfaces/). This is Konrad Polthier's modern version of Felix Klein's famous set of geometric models. Correspondingly, a modern interactive version of Euclid is provided by *Cinderella* (a software tool which is largely comparable with *The Geometer's Sketchpad;* the latter is discussed in detail in Chapter 7 of this volume). Klein, like de Morgan, was equally influential as an educator and as a researcher.

Sylvester's theorem

Sylvester's theorem is worth mentioning because of its elegant visual proof, but also because of Sylvester's complex relationship to geometry: "The early study of Euclid made me a hater of geometry" (quoted in MacHale, 1993, p. 135). James Joseph Sylvester, who was the second president of the London Mathematical Society, may have hated Euclidean geometry, but discrete geometry (now much in fashion under the name 'computational geometry', offering another example of very useful pure mathematics) was different. His strong, emotional preference nicely illustrates how the aesthetic is involved in a mathematician's choice of fields.

Sylvester (1893) came up with the following conjecture, which he posed in *The Educational Times:*

THEOREM Given n non-collinear points in the plane, then there is always at least one (*elementary* or *proper*) line going through exactly two points of the set.

Sylvester's conjecture was, so it seems, forgotten for fifty years. It was first established – 'badly', in the sense that the proof is much more complicated – by T. Grünwald (Gallai) in 1933 (see editorial comment in Steinberg, 1944) and also by Paul Erdös. Erdös, an atheist, named 'the Book' the place where God keeps aesthetically perfect proofs. L. Kelly's proof (given below), which Erdös accepted into 'the Book', was actually published by Donald Coxeter (1948) in the *American Mathematical Monthly*. This is a fine example of how the archival record may rapidly get obscured.

PROOF Consider the point *closest* to a line it is not on and then suppose that line has three points on it (the horizontal line). The middle of those three points is clearly *closer* to the other line.

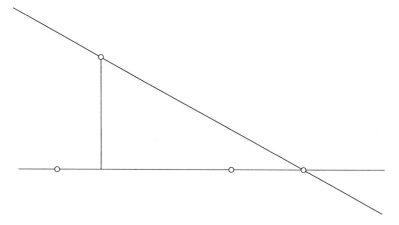

Figure 5. Kelly's proof from 'the Book'

As with Apostol's proof of the irrationality of $\sqrt{2}$, we can see the power of the right *minimal configuration*. Aesthetic appeal often comes from having this characteristic: that is, its appeal stems from being able to reason about an unknown number of objects by identifying a restricted view that captures all the possibilities. This is a process that is not so very different from that powerful method of proof known as mathematical induction.

Another example worth mentioning in this context (one that belongs in 'the Book') is Niven's (1947) marvellous (simple and short), half-page proof that π is *irrational* (see: www.cecm.sfu.ca/personal/jborwein/pi.pdf).

Partitions and patterns

Another subject that can be made highly accessible through experimental methods is additive number theory, especially *partition theory*. The number of *additive partitions* of n, p(n), is generated by the following equation:

$$f(q) := \prod_{k \geq 1} \frac{1}{1 - q^k} = \sum_{n=0}^{\infty} p(n)q^n.$$

Thus, p(5) = 7, since:

$$5 = 4 + 1 = 3 + 2 = 3 + 1 + 1 = 2 + 2 + 1$$

$$= 2 + 1 + 1 + 1 = 1 + 1 + 1 + 1 + 1$$

QUESTION How hard is p(n) to compute? Consider this question as it might apply in 1900 (for Major MacMahon, the father of our modern combinatorial analysis) and in 2000 (for *Maple*).

ANSWER Seconds for *Maple,* months for MacMahon. It is interesting to ask if development of the beautiful asymptotic analysis of partitions by Hardy, Ramanujan and others would have been helped or impeded by such facile computation.

Ex-post-facto algorithmic analysis can be used to facilitate independent student discovery of *Euler's pentagonal number theorem.*

$$\prod_{k \geq 1}(1 - q^k) = \sum_{n=-\infty}^{\infty} (-1)^n q^{(3n+1)n/2}.$$

Ramanujan used MacMahon's table of p(n) to intuit remarkable and deep congruences, such as:

p($5n + 4$) ≡ 0 (mod 5),

p($7n + 5$) ≡ 0 (mod 7),

and

p($11n + 6$) ≡ 0 (mod 11)

from data such as:

$$
\begin{aligned}
f(q) \;=\; & 1 + q + 2q^2 + 3q^3 + 5q^4 + 7q^5 + 11q^6 + 15q^7 + 22q^8 + 30q^9 \\
& + 42q^{10} + 56q^{11} + 77q^{12} + 101q^{13} + 135q^{14} + 176q^{15} + 231q^{16} \\
& + 297q^{17} + 385q^{18} + 490q^{19} + 627q^{20} + 792q^{21} + 1002q^{22} \\
& + 1255q^{23} + \dots
\end{aligned}
$$

Nowadays, if introspection fails, we can recognise the *pentagonal numbers* occurring above in Sloane and Plouffe's on-line *Encyclopaedia of Integer Sequences* (see: www.research.att.com/personal/njas/sequences/eisonline.html). Here, we see a very fine example of *Mathematics: the Science of Patterns,* which is the title of Keith Devlin's (1994) book. And much more may similarly be done.

Some Concluding Discussion

In recent years, there have been revolutionary advances in cognitive science – advances that have a profound bearing on our understanding of mathematics. (More serious curricular insights should come from neuro-biology – see Dehaene *et al.*, 1999.) Perhaps the most profound of these new insights are the following, presented in Lakoff and Nuñez (2000).

1. *The embodiment of mind.* The detailed nature of our bodies, our brains and our everyday functioning in the world structures human concepts and human reason. This includes mathematical concepts and mathematical reason. (See also Chapter 6.)
2. *The cognitive unconscious.* Most thought is unconscious – not repressed in the Freudian sense, but simply inaccessible to direct conscious introspection. We cannot look directly at our conceptual systems and at our low-level thought processes. This includes most mathematical thought.
3. *Metaphorical thought.* For the most part, human beings conceptualise abstract concepts in concrete terms, using ideas and modes of reasoning grounded in sensori-motor systems. The mechanism by which the abstract is comprehended in terms of the concept is called *conceptual metaphor.* Mathematical thought also makes use of conceptual metaphor: for instance, when we conceptualise numbers as points on a line.

Lakoff and Nuñez subsequently observe:

> What is particularly ironic about this is it follows from the empirical study of numbers as a product of mind that it is natural for people to believe that numbers are not a product of mind! (p. 81)

I find their general mathematical schema pretty persuasive but their specific accounting of mathematics forced and unconvincing (see also Schiralli and Sinclair, 2003). Compare this with a more traditional view, one that I most certainly espouse:

> The price of metaphor is eternal vigilance. (Arturo Rosenblueth and Norbert Wiener, in Lewontin, 2001, p. 1264)

Form follows function

> The waves of the sea, the little ripples on the shore, the sweeping curve of the sandy bay between the headlands, the outline of the hills, the shape of the clouds, all these are so many riddles of form, so many problems of morphology, and all of them the physicist can more or less easily read and adequately solve [...] (Thompson, 1917/1968, p. 10)

A century after biology started to think physically, how will mathematical thought patterns change?

> The idea that we could make biology mathematical, I think, perhaps is not working, but what is happening, strangely enough, is that maybe mathematics will become biological! (Chaitin, 2002)

To appreciate Greg Chaitin's comment, one has only to consider the metaphorical or actual origin of current 'hot topics' in mathematics research: simulated annealing ('protein folding'); genetic algorithms ('scheduling problems'); neural networks ('training computers'); DNA computation ('travelling salesman problems'); quantum computing ('sorting algorithms').

Humanistic philosophy of mathematics

However extreme the current paradigm shifts are and whatever the outcome of these discourses, mathematics is and will remain a uniquely human undertaking. Indeed, Reuben Hersh's (1995) full argument for a humanist philosophy of mathematics, as paraphrased below, becomes all the more convincing in this setting.

1. *Mathematics is human.* It is part of and fits into human culture. It does not match Frege's concept of an abstract, timeless, tenseless and objective reality (see Resnik, 1980, and Chapter 8). It shares important features with the other humanities, including an appreciation for the role of intuition and an understanding of the value judgements that help determine what is investigated, how it is investigated and why.

2. *Mathematical knowledge is fallible.* As in science, mathematics can advance by making mistakes and then correcting or even re-correcting them. The 'fallibism' of mathematics is brilliantly argued in Imre Lakatos's (1976) *Proofs and Refutations.*

3. *There are different versions of proof or rigour.* Standards of rigour can vary depending on time, place and other things. Using computers in formal proofs, exemplified by the computer-assisted proof of the four-colour theorem in 1977, is just one example of an emerging, non-traditional standard of rigour.

4. *Aristotelian logic is not always necessarily the best way of deciding.* Empirical evidence, numerical experimentation and probabilistic proof can all help us decide what to believe in mathematics.

5. *Mathematical objects are a special variety of a social–cultural–historical object.* Contrary to the assertions of certain post-modern detractors, mathematics cannot be dismissed as merely a new form of literature or religion. Nevertheless, many mathematical objects can be seen as shared ideas, like *Moby Dick* in literature or the Immaculate Conception in religion.

The recognition that 'quasi-intuitive' methods may be used to gain good mathematical insight can dramatically assist in the learning and discovery of mathematics. Aesthetic and intuitive impulses are shot through our subject and honest mathematicians will acknowledge their role.

Some Final Observations

> When we have before us, for instance, a fine map, in which the line
> of coast, now rocky, now sandy, is clearly indicated, together with
> the windings of the rivers, the elevations of the land, and the distri-
> bution of the population, we have the simultaneous suggestion of so
> many facts, the sense of mastery over so much reality, that we gaze
> at it with delight, and need no practical motive to keep us studying
> it, perhaps for hours together. A map is not naturally thought of as
> an æsthetic object; it is too exclusively expressive. (Santayana, 1896/
> 1910, p. 209)

This Santayana quotation was my earliest, and still favourite, encounter with
aesthetic philosophy. It may be old fashioned and un-deconstructed in tone,
but to me it rings true. He went on:

> And yet, let the tints of it be a little subtle, let the lines be a little del-
> icate, and the masses of land and sea somewhat balanced, and we
> really have a beautiful thing; a thing the charm of which consists
> almost entirely in its meaning, but which nevertheless pleases us in
> the same way as a picture or a graphic symbol might please. Give
> the symbol a little intrinsic worth of form, line, and color, and it
> attracts like a magnet all the values of the things it is known to sym-
> bolize. It becomes beautiful in its expressiveness. (p. 210)

However, in conclusion, and to avoid possible accusations of mawkishness
at the close, I also quote Jerry Fodor (1985):

> It is, no doubt, important to attend to the eternally beautiful and to
> believe the eternally true. But it is more important not to be eaten.
> (p. 4)

Notes

[1] This quotation is commonly attributed to Gauss, but it has proven remarkably
resistant to being tracked down. Arber, the citation I give here, a philosopher of
biology, acknowledges in a footnote (p. 47) that, "the present writer has been unable
to trace this dictum to its original source". Interestingly, even the St. Andrews history of mathematics site cites Arber. See also Dunnington (1955/2004).

[2] In *Brouwer's Cambridge Lectures on Intuitionism,* the editor van Dalen (1981,
p. 95) comments in a footnote:

> 3. The first use of undecidable properties of effectively presented
>   objects (such as the decimal expansion of π) occurs in Brouwer
>   (1908 [/1975]).

CHAPTER 2
Beauty and Truth in Mathematics

Doris Schattschneider

"That's beautiful!" is the unsolicited exclamation. The response is not to a painting, a breathtaking view or a flawless musical performance, but rather to a mathematical statement or a mathematical proof. What brings such aesthetic pleasure to a mathematician or to those who wish to appreciate mathematics and engage in it?

Beautiful Statements

A simple, yet profound statement can evoke awe. Perhaps one of the most surprising of all mathematical truths is:

$$e^{i\pi} + 1 = 0.$$

Here, in one incredibly spare equation, five of the most important numbers are related. That's beautiful!

Another mathematical truth, discovered by Archimedes (c. 240BCE), is a geometric rival to the numerical epigram above. Figure 1 shows a 1 by 2 rectangle and, on its base, a semi-circle and an isosceles triangle are inscribed.

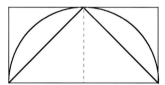

Figure 1: Archimedes's discovery

From elementary calculations, it can be seen that the three areas are in the following ratios to one another:

area of triangle : area of semi-circle : area of rectangle
1 : $\pi/2$: 2

If these three figures are rotated about the dashed vertical axis in the figure, then a cone, a hemisphere and a cylinder are swept out simultaneously, all having the same radius and height. Archimedes discovered this beautiful relationship among their volumes:

volume of cone : volume of hemisphere : volume of cylinder
1   :   2   :   3

Archimedes proved this relationship in his treatise *On the Sphere and the Cylinder*. It is claimed he held this relationship to be so beautiful that he asked that the diagram of a sphere inscribed in a cylinder be carved on his gravestone. Howard Eves (1980) presents Archimedes's proof of the relationship, and recounts as well the often-told tale of how Archimedes died at the hands of a Roman soldier who did not share his passion for mathematics.

Perhaps one of the (if not *the*) most well-known (also well-historied and well-proved) mathematical statements is what the Western world calls:

> **The Pythagorean theorem** The sum of the (areas of the) squares on the legs of a right triangle equals the (area of the) square on the hypotenuse.

Surely, the first time this relationship was discovered (and the many times it has been rediscovered), the discoverer must have been awe-struck. Today, it is just one of many geometric 'facts', rarely discovered, but rather memorized as an algebraic equation, often without hypotheses:

$$a^2 + b^2 = c^2$$

In his book, *Geometry Civilized*, J. L. Heilbron (1998) discusses some of this theorem's history and notes that knowledge of this beautiful theorem has been considered, by many, to be the mark of a civilized person. Some have even proposed having what Heilbron and others have called Euclid's 'windmill' diagram (Figure 2) engraved in huge proportions on the landscape to be seen from outer space. Perhaps it is not only the theorem's beauty, but its incredible usefulness that has made it the object of enduring admiration (see also Valens, 1964).

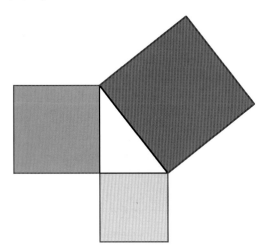

Figure 2: The Pythagorean theorem as an icon

Beautiful Proofs

A mathematical *proof,* more so than a mathematical statement or fact, is likely to be labeled 'beautiful' by a mathematician. What makes a proof beautiful? Here are some of the characteristics that most mathematicians would agree on:

- *elegance* – it is spare, cutting right to the essential idea;
- *ingenuity* – it has an unexpected idea, a surprising twist;
- *insight* – it offers a revelation as to *why* the statement is true, it provides an *Aha!* ;
- *connections* – it enlightens a larger picture or encompasses many areas;
- *paradigm* – it provides a fruitful heuristic with wide application.

To me, a proof is beautiful if it really catches my attention and, most of all, if it is one whose essence I will never forget. I have collected several proofs in this chapter that I feel embody one or more of the characteristics in the above list. Many can be shared with high-school students—it is not necessary to have research-level mathematics to encounter beautiful proofs.

I begin with one of the oldest recorded proofs of the *hsuan-thu* (the Pythagorean theorem) that appears in an ancient Chinese mathematical text entitled *Chou pei suan ching* (*The Arithmetical Classic of the Gnomon and the Circular Paths of Heaven,* possibly pre-third century BCE)—see Swetz and Kao (1980). Although adapted to today's use of letter symbols, it perfectly illustrates two paradigms of proof: *visualization,* with its use of diagram and color, and *finite dissection and reassembly,* preserving areas. The Chinese diagram shown in Figure 3 is the same as one given by the twelfth-century Indian scholar Bhāskara, whose one-word injunction *Behold!* recorded his sense of awe.

In the Chinese proof, regions are colored red and yellow. Figure 3 uses shades of gray to represent those colors. There are eight right triangles in the diagram, all congruent to triangle DEF. Figure 3 displays the proof that the square (ADFK) on the hypotenuse of right triangle DEF has the same area as the sum of squares that can be built on its legs (ABCH + CEFG).

Euclid's proof of the Pythagorean theorem (Book I, Prop. 47) is, by comparison, more complicated. He used the 'windmill' diagram and had the ingenious idea of adding strategically chosen auxiliary lines to dissect the square on the hypotenuse into two rectangles whose respective areas are equal to the areas of the squares on the adjacent legs. Figure 4 shows Euclid's famous diagram for his proof.

Euclid's proof is not only ingenious, but is beautiful for another reason —the argument extends, almost without change, to proving Pappus's Theorem, from Book IV of his *Collection* (fourth century CE), which has the Pythagorean theorem as a mere special case. The diagram for the theorem of Pappus is shown in Figure 5: my statement of the theorem refers to the labeling in that diagram.

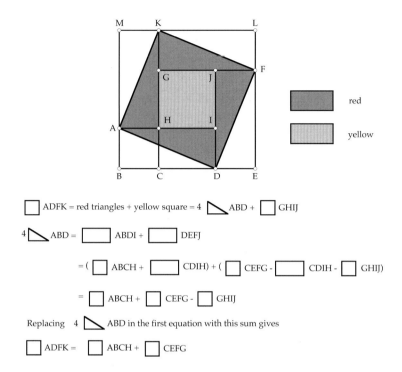

ADFK = red triangles + yellow square = 4 ▲ ABD + ☐ GHIJ

4 ▲ ABD = ☐ ABDI + ☐ DEFJ

= (☐ ABCH + ☐ CDIH) + (☐ CEFG - ☐ CDIH - ☐ GHIJ)

= ☐ ABCH + ☐ CEFG - ☐ GHIJ

Replacing   4 ▲ ABD in the first equation with this sum gives

☐ ADFK = ☐ ABCH + ☐ CEFG

Figure 3: An ancient Chinese proof of the Pythagorean theorem

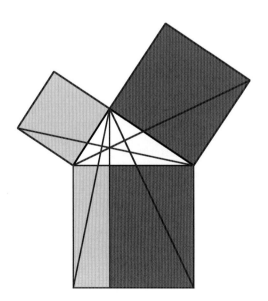

Figure 4: Euclid's proof of the Pythagorean theorem, c. 300BCE

Pappus's theorem Let ABC be any triangle with parallelograms ACDE and ABFG constructed externally on the sides AC and AB. Let the rays DE and FG meet in point H, and construct BJ and CK equal and parallel to HA. Then the sum of the areas of the parallelograms ACDE and ABFG equals the area of the parallelogram BCKJ.

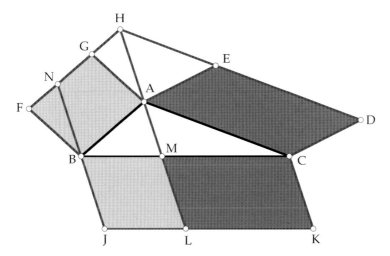

Figure 5: Pappus's theorem, Book IV, The Collection, *fourth century* CE

Pappus had the vision to see beyond a right triangle and the insight to find the right generalization of the Pythagorean theorem. Perhaps his theorem and its proof can serve as an example of a proof that makes connections and illuminates a larger picture. In her poem 'Poet as mathematician', Lillian Morrison (1979, p. 45) captures the essence of this mathematical beauty:

> Having perceived the connexions, he seeks
> the proof, the clean revelation in its
>
> simplest form, never doubting that somewhere
> waiting in the chaos, is the unique
>
> elegance, the precise, airy structure,
> defined, swift-lined, and indestructible.

Another of Euclid's proofs, this time from number theory, is a masterpiece of elegance. Euclid proved that the primes never run out—that is, *given any finite collection of primes, there is always one more*. His brilliant idea, which reduces the proof to a few lines, is to take the product of the given collection of primes and add 1. That new number (which may be very large) is not divisible by any of the primes in the collection. But every number greater than 1 can be factored into a product of primes, so there must be a prime *not* in the original collection. For example, if we begin with the six primes 2, 3, 5, 7, 11, 13, then the new number that is produced by this construction is:

$$2 \times 3 \times 5 \times 7 \times 11 \times 13 + 1 = 30{,}031$$

which is not divisible by any of 2, 3, 5, 7, 11 or 13. In this case, the number obtained is not itself prime (as many students tend to think it *always* must be): in fact, 59 is the smallest prime that divides 30,031 (– 59 × 509).

Interestingly, Euclid's *reductio ad absurdum* proof that the set of primes cannot be finite is actually a generic one. It only shows the result for a collection of three primes, using the above construction in this particular case to argue for the necessary existence of a fourth not included in the list. There is not even a generalizing remark to the effect that 'for any other number of primes the argument runs likewise'.

My favorite visual proofs are ones that capture the whole idea of an assertion—they are like a haiku, a poem that is so spare yet searing that it can leave you breathless. For me, the best example is George Pólya's proof of:

> **The arithmetic–geometric mean inequality** For any two positive
> numbers, *a* and *b*,
>
> $$\sqrt{ab} \leq \frac{a+b}{2}$$

The diagram, which unequivocally demonstrates the inequality, is given below in Figure 6. (The circle, with just three lines, appears like a Japanese crest.) A circle is constructed on a diameter of length *a* + *b*. A segment perpendicular to the diameter is constructed from the point that separates length *a* from length *b* on the diameter, and joins that point to the circle. A simple calculation shows that this segment has length $\sqrt{ab}$, while the radius of the circle has length $\frac{a+b}{2}$. If more explanation is needed, the inscribed (necessarily right) triangle at the right in Figure 6 can be drawn and a simple argument using similar triangles will show that the segment's length is $\sqrt{ab}$. (In fact, this was the Greek construction method for the length $\sqrt{ab}$.) Not only does this diagram make clear *why* the inequality is true, but it also demonstrates that equality between the two means occurs only when *a* = *b*. The diagram captures the essence of the inequality in a minimal and graceful manner; it is a picture I shall never forget.

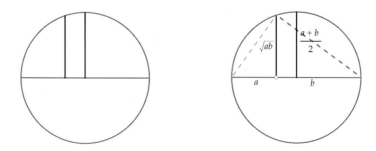

Figure 6: Pólya's proof of the arithmetic–geometric mean inequality

A haiku is not only spare, but must have exact structure: five syllables, seven syllables, five syllables. I was amazed to find that when I read the inequality in this theorem, either with the symbols in the algebraic form above or using descriptive words, it fitted the form of a haiku.

Two mean haikus

Square root of *a*, *b*   Geometric mean
is less than or equal to   is less than or equal to
av'rage of *a*, *b*.   arithmetic mean.

A real haiku is one that not only fits the beat pattern, but conveys its message without ever saying it explicitly. Pólya's diagram (minus the letters) is in the spirit of a true haiku.

Paradigms of Technique

Proofs that are paradigms of technique, that show an unconventional method that can be tried in another (perhaps similar) situation, evoke admiration. On encountering one of these, the reaction is often something like, "I'd never have thought of that". The fact that the technique is fruitful, and not just an isolated idiosyncrasy, is what makes it even more memorable. (Unfortunately, many of these paradigms are presented in texts without any fanfare, as if they were standard methods of argument. Yet they do not arise out of the body of mathematical logic—they are the fruit of inspired flashes of insight that have been absorbed into the larger body of mathematical argument.)

I have already presented examples of two paradigms of technique: finite dissection and reassembly (the Chinese proof of the Pythagorean theorem) and visualization or 'geometrization' (Pólya's proof of the arithmetic–geometric mean inequality). Some other paradigms are:

- complementarity (duality);
- the Fubini principle (counting twice);
- the pigeon-hole principle (the Dirichlet principle);
- parity;
- transformation;
- symmetry;
- patterns.

The examples I have chosen to illustrate these may seem well-worn to some, but for someone who has never seen such arguments, they can evoke enthusiasm and admiration.

Complementarity (duality)

This principle is encapsulated in many combinatorial proofs. For example, if you have a set of n elements, and you choose r of them, you have automatically designated $n - r$ elements as "not chosen". Each choice of r defines a complementary choice of $n - r$; this establishes a one-to-one correspondence between complementary subsets of the given set. This observation actually constitutes a proof of the following theorem:

> There are exactly as many r-element subsets of a set of n elements as there are $(n - r)$-element subsets.

Using $C(n, r)$ to denote the number of subsets of r elements chosen from an n-element set, the theorem says $C(n, r) = C(n, n - r)$. Although algebraic expressions for $C(n, r)$ and $C(n, n - r)$ can also be manipulated to verify the truth of this theorem, the obvious truth of the complementarity argument is more convincing (and surprising, since it involves no computation at all).

Using the same notation, another theorem states:

> If $n > 1$, then $C(n, r) = C(n - 1, r) + C(n - 1, r - 1)$.

A complementarity argument also makes this clear: choose one element, x_0, from the given set of n elements. Count all the subsets of r elements that do not contain x_0; there are $C(n - 1, r)$ of these (since, by withholding x_0, there are only $n - 1$ elements from which to choose r). Then count all the subsets of r elements that do contain x_0; there are $C(n - 1, r - 1)$ of these (since, as one choice, x_0, is prescribed, there are only $r - 1$ choices left to make from $n - 1$ elements). As there is no overlap of the subsets counted, the identity is proved.

The Fubini principle (counting twice)

Sherman Stein (1979), noting the Fubini theorem from calculus allowing interchanging the order of integration of double integrals, gives this name to the more general principle that it does not matter in which order you sum. This means that when you add up a collection of numbers in two different ways, and then equate the results, you often discover unexpected relationships and formulas. Stein gives several examples of this paradigm; more can be found in Schattschneider (1991).

A related technique is to add numbers in such a way as to get twice the desired sum and then divide by 2. The most well-known example of a proof using this principle is the often-told tale of how the young Gauss found the sum of the first 100 consecutive integers. His solution works for any positive integer n, not just 100, and goes like this. Write the sum from 1 to n horizontally, then write the same sum backwards from n to 1 and arrange the two sums vertically. All the (vertical) pairs add to $n + 1$, and there are n of these pairs, so twice the sum from 1 to n equals $n(n + 1)$. Divide by 2 and you have the sum from 1 to n. What ingenuity!

$$
\begin{array}{ccccccccc}
1 & + & 2 & + & 3 & + & \cdots & + & n \\
\underline{n} & + & \underline{(n-1)} & + & \underline{(n-2)} & + & \cdots & + & \underline{1} \\
(n+1) & + & (n+1) & + & (n+1) & + & \cdots & + & (n+1)
\end{array}
$$

Therefore, $\quad 1 \quad + \quad 2 \quad + \quad 3 \quad + \quad \cdots \quad n = \dfrac{n(n+1)}{2}$

A totally visual, generic proof of this sum formula, using the same technique of combining two equal sums and then taking half, is shown in Figure 7. Here, the unit is represented by a square of area 1 and the area of the dark stair-steps represents the sum from 1 to n. Two of these stair-steps combine to form an $n \times (n + 1)$ rectangle; so half of that area is $\frac{n(n + 1)}{2}$.

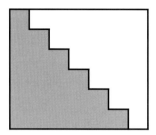

Figure 7: A visual proof of the formula for the sum from 1 to n

Surprisingly, Gauss's technique extends to summing *any* arithmetic sequence. Here the sequence's first term is a, the difference between consecutive terms is d and there are $k + 1$ terms in the sequence (d is added to successive terms k times to get from the first to the last term). As was done with the sum from 1 to n (in which $a = 1$, $d = 1$, and $n = k + 1$), you can line up the two sums, add vertically and divide by 2 to obtain the following result:

The sum of any arithmetic sequence is:

$$\frac{(\text{first term} + \text{last term}) \times (\text{number of terms})}{2}.$$

A visual proof of this result can be seen in Figure 8; it, too, is a direct generalization of the generic image for the simpler case.

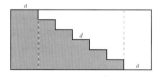

Figure 8: A visual proof of the formula for the sum of any arithmetic sequence

The pigeon-hole principle (the Dirichlet principle)

The pigeon-hole principle states the obvious: if there are more pigeons than pigeon-holes, then (provided all the pigeons are in holes) at least one hole contains at least two pigeons. There are, of course, generalizations of this innocuous observation. Who would think that this could be the basis for a mathematical proof? One of the most elegant applications of this principle was made by Louis Pósa, when he was only eleven years old. Even mathematician Paul Erdös was impressed. Honsberger (1973) recounts the story as told by Erdös:

> I met him [Pósa] before he was 12 years old. When I returned from the United States in the summer of 1959 I was told about a little boy whose mother was a mathematician and who knew quite a bit about high school mathematics. I was very interested and the next day I had lunch with him. While Pósa was eating his soup I asked him the following question: Prove that if you have $n + 1$ positive integers less than or equal to $2n$, some pair of them are relatively prime. It is quite easy to see that the claim is not true of just n such numbers because no two of the n even numbers up to $2n$ are relatively prime. Actually I discovered this simple result some years ago but it took me about ten minutes to find the really simple proof. Pósa sat there eating his soup, and then after a half a minute or so he said "If you have $n + 1$ positive integers less than or equal to $2n$, some two of them will have to be consecutive and thus relatively prime." Needless to say, I was very much impressed, and I venture to class this on the same level as Gauss' summation of the positive integers up to 100 when he was just 7 years old. (pp. 10-11)

What Erdös leaves out (because it is obvious to him) is the reason there must be two consecutive integers among the $n + 1$. This is because there are a maximum of n non-consecutive 'pigeon-holes' among $2n$ of them in a line and so, by the pigeon-hole principle, placing $n + 1$ numbers in the $2n$ slots would force (at least) two of them to be in consecutive holes. Pósa's one-line proof is indeed impressive! Some other delightful applications of the pigeon-hole principle can be found in Rebman (1979) and Stein (1979).

Parity

Simply noting whether a number is even or odd would not seem to be particularly useful as a mathematical tool for proof. Yet, in the right circumstance, it is all that is needed for a convincing argument. Honsberger (1973, pp. 64-65) reports a particularly nice application in proving the following theorem:

> The number of divisors of a positive integer is odd if and only if n is a perfect square.

To see this, simply note that unequal divisors come in pairs: $n = ab$ where $a < \sqrt{n}$ and $b > \sqrt{n}$, so these pairs account for all divisors of n unless $n = a^2$ for some integer a, in which case (and only in this case) there is an odd number of divisors.

Although number theory is a natural area in which you would expect parity arguments to arise, parity is also a powerful tool in various problems that can be translated into tiling or coloring problems (especially tiling or covering an $m \times n$ chessboard).

Transformation

Often a problem will fail to yield to direct attack or, if it does, the work is lengthy and clumsy, yielding no insight. To transform the problem to another setting is often the means required for illuminating the picture, as well as for producing a simple solution. Of course, how and when to transform are never obvious and it sometimes takes a stroke of brilliance to effect this technique. It can be as simple as transforming an algebraic problem into a geometric one (as seen in earlier examples), altering a symbolic expression to a story interpretation (counting problems) or applying a formal transformation to a given circumstance.

Here, an invertible function transfers the given problem into another space, but in such a manner that the characteristics essential to the problem are preserved (remain invariant). In the new setting, the problem is solved and then the inverse transformation carries the solution back to the original setting. Eves (1972) calls this technique 'transform—solve—invert'. It can be especially fruitful in proofs of geometric statements, but is also effective in algebraic or other settings. For example, geometric transformations can take a configuration to an ideal special case or to a setting in which deriving a proof is far simpler. Group representations can transform elements of groups into matrices or into other forms that are more amenable to computation or argument.

Theorems in geometry on concurrence, collinearity, parallelism and also tangency are all candidates for this technique: for example, concurrence of medians is easily proved in an equilateral triangle. To prove this concurrence in any triangle, you need only apply a transformation that sends the given triangle onto an equilateral one, while preserving lines (and hence concurrence) and mid-points—an affine transformation will do.

An elegant solution to a more complicated problem of Jakob Steiner also yields to this approach.

> **Steiner's circle problem** Given a circle, with a second circle in its interior, is it (or when is it) possible to construct a chain of circles between the two given ones, so that each circle in the chain is tangent both to the given circles and to the two adjacent circles in the chain?

When the two given circles are concentric, the problem is easy to solve—with some elementary calculations, you can give precise necessary and sufficient conditions on the radii of the two given circles to make the construction described. But the question of how to tackle the general problem seems hopeless.

The surprising solution is to use the fact that the two given circles can be transformed into two concentric ones by an inversion in a (third) circle and this inversion sends every circle to another one and preserves tangency. If the inverted versions of the given circles (now concentric) fit the requirements that the desired chain of tangent circles can be constructed, then a chain can be constructed in the given circles—just construct the chain in the concentric circles and then invert it (back) to produce a chain in the given circles. And, of course, if the concentric circles do not fit the requirements to allow a chain of tangent circles, then neither do the given circles. Figure 9 gives an illustration—here, the original circles (with their chains of tangent circles) are on the right, and their concentric images under inversion in a dashed circle are the ones on the left.

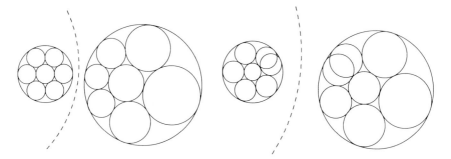

Figure 9: Inversion provides a solution to Steiner's circle problem

Symmetry

A well-known and often-used paradigm not only for mathematics, but for all science, is symmetry (for example, see Weyl, 1952; Curtin, 1982). When expressions, statements or configurations possess symmetry (such as symmetric polynomials, cyclic permutations or geometric figures invariant under reflection or rotation, for example), this symmetry can often be utilized to produce a streamlined proof. Indeed, symmetry may be the root cause of a particular relationship. The Platonic solids, or regular polyhedra, are a personification of symmetry: each has congruent regular polygonal faces whose vertices meet to form congruent solid corners—turning the solid makes no difference, it looks the same no matter from which face you view it. (This can be described more precisely in terms of symmetry groups: each corner can be mapped onto any other corner by a rotation or reflection that leaves the solid invariant.) This high symmetry accounts for the fact that there are only five regular polyhedra. Since every corner must be the same, Euclid, in his proof of this result, only needed to consider how many congruent regular polygons could come together to make a solid corner. For triangles, only three, four or five (tetrahedron, octahedron, icosahedron respectively); for squares, only three (cube); for pentagons, only three (dodecahedron): there are no others. The *Platonic Solids* video (Schattschneider and Fetter, 1991) provides a delightful animation of this argument.

The high symmetry of each Platonic solid can also be effectively utilized to find relationships among the numbers of its edges (e), faces (f), and vertices (v). For example, on the dodecahedron, we can use the fact that every face has five edges and each edge is common to exactly two faces. Counting all the edges around all the faces (note each edge is counted by two faces), we obtain $5f = 2e$.

Also, three faces meet at each vertex of the dodecahedron and each face is surrounded by five vertices. Counting all the vertices around all the faces, we obtain $5f = 3v$. Although we could, at this point, deduce that $2e = 3v$, this can also be obtained by another count. Each vertex has three edges that meet there and each edge has exactly two vertices as endpoints. Counting all the edges that meet at all the vertices, we obtain $3v = 2e$. This kind of counting can be adapted to any polyhedron having high symmetry (such as a semi-regular polyhedron). More applications of symmetry can be found in Schattschneider and Fetter (1991).

For centuries, symmetry (and, in the last century, symmetry groups and their actions) has provided useful techniques in mathematics and science, both in proving results and in modeling structure (such as that of molecules and crystals). Symmetry is beautiful and it is effective, but this beauty is seductive. Here is where I wish to address the two key words in my title, inspired by Keats's (1819) famous line 'Beauty is truth, truth beauty', from his poem *Ode on a Grecian Urn*.

'Beauty is truth' is the seductive claim—more than one mathematician or scientist has fallen prey to this seduction. Johannes Kepler's polyhedral model of the universe is a case in point. He was so convinced that the perfection of geometric symmetry was the key to understanding the orbits of the planets that he constructed a model in which each planet (that was known at the time) traveled its path in a spherical shell, its place in the model dictated by the strategic nesting of the five Platonic solids.

Cromwell (1997) notes that:

> [Kepler] was motivated by the desire to expose [the universe's] mathematical design, to reveal the plan which the Creator had used in its construction. He followed in the Pythagorean tradition and believed that such a plan would be expressible in harmonious geometrical relationships reflecting the decision of the Architect. He did not believe that the polyhedra and crystal spheres actually existed in space; he thought of them more as an invisible skeleton, as part of the perfect design by which each planet was allotted its own region of space. The illusory and fallacious nature of the planetary model was shown up by the discovery of new planets after Kepler's death. [...] Just as Kepler admired the regularity of the Platonic solids and was attracted by the idea that nature must be constructed around such elegant forms, so the modern physicist idolises symmetry. (p. 148)

In fact, symmetry can be tyranny. By assuming that nature is defined by symmetry, scientists have restricted their methods of analysis of structure by looking for symmetry and applying symmetry groups. They have believed their models to be the truth and have not been open to looking for other ways to understand structure. Only in the last twenty years has this tyranny of symmetry been questioned, as the discovery of quasicrystals in 1984 destroyed the traditional definition of crystal and new paradigms such as self-similarity and repetitiveness of patterns have gained interest. In mathematics and science, beauty may not be truth.

Patterns

Another fruitful paradigm in solving mathematical problems is searching for (and recognizing) patterns. (This theme is also the cental focus of Chapter 5.) By looking at many specific cases of a general statement, by producing computer calculations or pictures of randomized examples of a general conjecture, or by carrying out a search for patterns among a vast array of data, a mathematician can often glean not only a pattern, but also insight into why a conjecture may be or could not be true. Often it is the evidence of pattern that convinces the mathematician of a truth. Only after this conviction is established does the search for a proof become serious. Michael de Villiers (1999) comments on this aspect of exploration and observation of (geometric) patterns as a precursor to proof.

Pattern alone is never a proof, although the exercise in finding patterns (especially numerical ones) is gaining such popularity in teaching mathematics that students seem to believe all the more that a pattern holding true for ten cases or so constitutes a proof that the pattern always holds (see Hewitt, 1992). Richard Guy (1988) once coined the phrase 'The strong law of small numbers' and that is what seems to be prevalent in the pointless search for patterns. Pattern, too, can be tyranny—for the familiar patterns we look for may not be the ones that lead to truth. The dark movie *Pi* provided a sad caricature of a man so obsessed with finding a pattern (in the digits of π) that it indeed destroyed him.

The Aesthetic of *Doing* Mathematics

For most of us, mathematics means fairly routine calculations and straightforward deductions from given premises, using various bits of mathematical knowledge to solve problems. But what grabs a mathematician? What makes a problem 'interesting'? Why is a mathematician willing to spend hours, days, months or even years trying to solve a mystery? Each one may have his or her own answer, but all wish to be able to make connections, to find an epiphany of understanding, to feel the intellectual and emotional high of accomplishment or the smaller satisfaction of making some inroads into an apparently intractable problem.

Artist Sarah Stengle, as the daughter of a mathematician, grew up knowing first-hand what mathematicians do. She writes:

> Mathematical imagery is seen through a veil of cultural assumptions that mathematics is unemotional and pure, and that its texts are stylistically neutral. (2000, p. 161)

Stengle questions these assumptions, arguing:

> Both the research mathematician and the artist proceed by intuition, often aesthetically motivated, and both share a sense of discovery and achievement if and when the desired outcome is attained. [...] Intuition and a desire for comprehension and beauty can motivate a mathematician or an artist. A mathematical proof which is entirely correct but boring is second rate, just as a portrait can be an excellent likeness but artistically dull. [Both are lacking 'essence' or 'soul'.] Mathematicians and artists often use a similar vocabulary of a search led by intuition to describe their working process. Often the outcome is described in terms of discovery, meaning the outcome was not known beforehand but seemed to exist *a priori*. Both often have only a sense of the outcome rather than a knowledge of it, and follow their intuition to their goals, which they recognize only when they get there. "There" is where things "feel" resolved and complete. The mathematical discovery has to withstand the rigid demands of the discipline, while the artistic discovery is subject to constant reinterpretation and debate. [Mathematical arguments are also subject to constant reinterpretation and debate.] (pp. 161, 165)

Although mathematicians all hope to experience searing insight, yearn to produce a flawless gem of a proof (exhibiting many of those characteristics I have listed earlier) or dream of cracking a problem that has baffled the best minds, most of us experience much lesser satisfactions. Yet that is often enough. If this were not so, very little mathematics would ever get done.

A recent, best-selling book, *Uncle Petros and Goldbach's Conjecture* (Doxiadis, 2000), paints a picture of a mathematician, a failure in both his family's eyes and his own, but admired by his persistent nephew who strives to find out what drove his uncle to this present situation. He learns of the lure of an intractable unsolved problem, the dedication to solving it in total isolation (to the exclusion of other mathematical research, even refusing to publish partial results) and the self-branding of failure when the task was not accomplished. It is a caricature in some ways, but also a sad commentary on the harshness with which the mathematical world judges mathematical prowess and results. At the same time, it celebrates the tenacity shown by mathematicians who are drawn to a problem and, despite many setbacks, are sufficiently encouraged by small successes to continue their search for a proof.

I recently had a personal experience in proving a theorem and, when done, I knew that the proof failed all the tests of being beautiful. The theorem is one about tiling the plane with congruent polygons. It says, in essence, that *if, in the tiling, every tile is surrounded in exactly the same way, then any isometry that maps one particular tile onto another chosen tile will map the whole tiling onto itself* (every tile will land exactly on another tile.) The mathematician who conjectured the theorem, Nikolai Dolbilin, had proved (in an elegant way) a powerful theorem (called the Local Theorem for tilings) that took care of this assertion for all polygons that had no mirror or rotation symmetry. So the case of when the polygonal tile was symmetric remained to be proved. A known theorem (about the topological network of edges of a tiling, called *Laves nets*) implied that polygons satisfying the hypothesis of the assertion could have no more than six sides. The only way to prove the symmetric case seemed to be to look at all possible cases—symmetric triangles, quadrilaterals, pentagons and hexagons that filled the plane in a manner satisfying the hypothesis of the assertion.

And so I carried out (during a concentrated six-month period) a case-by-case verification of the theorem. It turned out that there were forty cases and, although along the way I could sometimes see I was repeating arguments (and so could consolidate some arguments into lemmas to be used more than once), an all-encompassing argument or seminal idea never emerged.

The final result was a proof of the theorem (see Schattschneider and Dolbilin, 1998), but there was no elegance or ingenuity, nor was there any insight. In fact, the result, proved for tilings of the Euclidean plane, is known to be false both for the hyperbolic plane and also for Euclidean three-space. But the proof does not illuminate what is the essential difference, what is special about two-dimensional Euclidean space that makes the result true. And, moreover, Dolbilin and I were interested in proving the result for tiles of *any* shape—even those of Escher: we both believe the result to be true for this most general case in the plane. But the arguments in my plodding proof rely on the properties of polygons—in particular, that the sum of the interior angles of any n-gon is $(n - 2) \times 180°$—and I feel the general proof would require far more complicated arguments than I had employed.

Yet, I received a measure of satisfaction in proving this result. I had believed it was true and now it was proven to be true. And, along the way, I saw some connections that I could not have seen without actually going through the process. Begin with a tile type (for example, a hexagon mirror-symmetric about a side bisector) and surround it completely with copies of itself to form its 'first corona' (and sometimes even surround that corona with such tiles as necessary to form a 'second corona'). Then employ the Local Theorem to obtain the verification of the conjecture for that case.

I used *The Geometer's Sketchpad* to construct each tile and build up these tilings: in so doing, I could deform tiles and tilings that were flexible. This process did bring insight and connections to other cases. Figure 10 shows one unexpected connection that the dynamic geometry software revealed. (*The Geometer's Sketchpad* is the focus of Chapter 7.) This particular hexagon had five equal sides and one free edge (the bottom). When I reduced that free edge to a point, turning it into an equilateral pentagon, I could see that this particular pentagonal case was merely a special case of the hexagonal tiling. I had proved the pentagonal case first and had no inkling at the time that it was related to a more general hexagonal tiling.

Another satisfaction was that, when done, I realized that the case-by-case attack had produced a complete catalog of isohedral tilings by symmetric

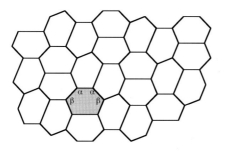 

Figure 10: A flexible tiling by symmetric hexagons having five equal sides and angle relation $\alpha + 2\beta = 2\pi$ *produces a related tiling by equilateral pentagons*

polygons—something that had not been done before. Moreover, the *Geometer's Sketchpad* constructions of these cases demonstrated, in a visceral way, the broad range of shapes of these tiles (convex and non-convex) and their tilings. (To view and manipulate them, see: mathforum.org/dynamic/one-corona/.)

Proofs of many theorems far more (mathematically) earth-shaking than this one have also been the result of long, arduous work (measured in years) with many false trails, some mistakes, many flashes of insight and connections and many small satisfactions along the way. They have often been guided by a program that outlined a framework of attack, so that pieces could be worked on (often by different mathematicians) and the proof chiseled out a bit at a time, until all the completed pieces fitted together. Just a few of these instances are: the classification of finite groups, the four-color theorem, Fermat's last theorem and Kepler's sphere-packing theorem. Those who completed these proofs are celebrated not for proofs that are concise or elegant, but for having the vision, the persistence and the stamina to complete the task and reach the goal. Truth can be beautiful in different ways.

And how are mathematicians rewarded for their sometimes pleasurable, often frustrating toil? Sherman Stein (1979) catches the wonder that keeps us at it:

> Frequently the reward for the answer to a question is the challenge of new questions. The mathematical unknown expands far more rapidly than it can be explored; it is full of galaxies of riddles as perplexing as the most peculiar star seen in a telescope. And the borders of this universe are restricted only by the extent of our curiosity and imagination. (p. 84)

I end with my own haiku, which perhaps sums up the emotions that we frequently encounter.

A mathematical haiku (after Dante)

Lightning strikes my mind
I see all, I have the proof!
And then I awake

CHAPTER 3
Experiencing Meanings in Geometry

David W. Henderson and Daina Taimina

What geometrician or arithmetician could fail to take pleasure in the symmetries, correspondences, and principles of order observed in visible things? Consider, even, the case of pictures: those seeing by the bodily sense the products of the art of painting do not see the one thing in the one only way; they are deeply stirred by recognizing in the objects depicted to the eyes the presentation of what lies in the idea, and so are called to recollection of the truth – the very experience out of which Love rises. (Plotinus, *The Enneads*, II.9.16; 1991, p. 129)

In mathematics, as in any scientific research, we find two tendencies present. On the one hand, the tendency toward *abstraction* seeks to crystallize the *logical* relations inherent in the maze of material that is being studied, and to correlate the material in a systematic and orderly manner. On the other hand, the tendency toward *intuitive understanding* fosters a more immediate grasp of the objects one studies, a live *rapport* with them, so to speak, which stresses the concrete meaning of their relations.

As to geometry, in particular, the abstract tendency has here led to the magnificent systematic theories of Algebraic Geometry, of Riemannian Geometry, and of Topology; these theories make extensive use of abstract reasoning and symbolic calculation in the sense of algebra. Notwithstanding this, it is still as true today as it ever was that intuitive understanding plays a major role in geometry. And such concrete intuition is of great value not only for the research worker, but also for anyone who wishes to study and appreciate the results of research in geometry. (David Hilbert, in Hilbert and Cohn-Vossen, 1932/1983, p. iii; *italics in original*)

It's a thing that non-mathematicians don't realize. Mathematics is actually an aesthetic subject almost entirely. (John Conway, in Spencer, 2001, p. 165)

The artist and scientist both live within and play active roles in constructing human mental and physical landscapes. That they should share structural intuitions is less surprising than inevitable. What is surprising and wonderful is how these intuitions have manifested themselves in the works of innovative artists and scientists in culturally apposite ways. (Kemp, 2000, p. 7)

The authors quoted above all stress the importance of the deep experience of meanings. It is these experiences in geometry (and indeed in all of mathematics, as well as in art and engineering) that we believe deserve to be called *aesthetic experiences*. Mathematics is a natural and deep part of human experience and experiences of meaning in mathematics should be accessible to everyone. Much of mathematics is not accessible through formal approaches except to those with specialized learning. However, through the use of non-formal experience and geometric imagery, many levels of meaning in mathematics can be opened up in a way that most people can experience and find intellectually challenging and stimulating.

A formal proof, as we normally conceive of it, is not the goal of mathematics—it is a tool, a means to an end. The goal is to understand meanings. Without understanding, we will never be satisfied—with understanding, we want to expand the meanings and to communicate them to others (see also Thurston, 1994). Many formal aspects of mathematics have now been mechanized and this mechanization is widely available on personal computers or even on hand-held calculators, but the experience of meaning in mathematics is still a human enterprise. Experiencing meanings is vital for anyone who wishes to understand mathematics or anyone wanting to understand something in their experience by means of the vehicle of mathematics. We observe in ourselves and in our students that such experiencing of meaning is, at its core, an aesthetic experience.

In this chapter, we recount some stories of our experience of meanings in geometry and art. David's story starts with art and ends with geometry, while Daina's story starts with geometry and ends with art. However, the bulk of what follows we both share.

David's Story: from Art to Mathematics

I have always loved geometry and have been thinking about geometric kinds of things ever since I was very young, as evidenced by a drawing I made when I was six years old (see Figure 1 overleaf).

The drawing is of a cat drawing a picture of a cat (who is presumably drawing a picture of a cat …). Notice the perspective from the point of view of the cat—for example, the drawing shows the underside of the table. I was already experiencing geometric meanings.

But I did not realize then that the geometry that I experienced was mathematics or even that it was called 'geometry'. I did not call it 'geometry' —I called it 'drawing' or 'design' or perhaps failed to call it anything at all and just did it. I did not like mathematics in school, because it seemed very dead to me—just memorizing techniques for computing things and I was not very good at memorizing. I especially did not like my high school geometry course, with its formal, two-column proofs.

However, I kept on doing geometry in various forms: in art classes, in carpentry, by woodcarving, when out exploring nature or by becoming

Figure 1: David's drawing (crayon on paper, 9" x 6")

involved in photography. This continued on into university where I became a joint physics and philosophy major, taking only those mathematics courses that were required for physics majors. I became absorbed by the geometry-based aspects of physics: mechanics, optics, electricity and magnetism, and relativity. On the other hand, my first mathematics research paper (on the geometry of Venn diagrams with more than four classes) evolved from a university course on the philosophy of logic. There were no geometry courses except for analytic geometry and linear algebra, which only lightly touched on anything geometric. So, it was not until my fourth and final year at the university that I switched into mathematics and I only did so then because I was finally convinced that the geometry that I loved really was a part of mathematics.

Since high school, I have never taken a course in geometry, because there were no geometry courses offered at the two universities I attended. Now I am a professional geometer and I started teaching an undergraduate Euclidean geometry course in the mid-1970s. My concern that both my students and I should experience meaning in the geometry quickly led me into conflict with traditional, formal approaches.

Daina's Story: from Mathematics to Art

I took a lot of geometry, both in grade school and at the university. But I only had a very few art lessons in school. From them, I developed the impression that I could not draw and that I had little artistic talent. But I liked geometry precisely for its aesthetic values. My mathematics teachers always paid a lot of attention to how we drew geometric diagrams; they encouraged Euclidean constructions with compass and straight-edge, but also supported the free-hand drawing of geometric figures, while insisting on accurate shapes and proportions. At university, besides other traditional geometry courses, I also took a course in descriptive geometry, as well as a short course on how to draw three-dimensional geometric diagrams—both of these latter courses contained a lot about perspective. I always enjoyed and excelled at the drawing aspects of geometry, but I did not think it had anything to do with art or aesthetic sensibilities.

When teaching the history of mathematics, I was particularly interested in the history of geometry and, because of my interest in art appreciation and art history, was happy to find so many connections between geometry and art. I was fascinated with the golden ratio, with the story of projective geometry arising from painters' perspective *prior* to it becoming a pure mathematical subject and with the considerable impact of mathematics on art in the twentieth century (for example, in cubism and, later, in the work of M. C. Escher). I was also teaching a university course on 'the psychology of mathematical thinking', which led me to wonder about all creative thinking.

I have had many students in my mathematics classes tell me that they were taking my class just to fulfill a distribution requirement. But they would also assert that they were no good at mathematics, because they are artists (poets, musicians, actors, painters) and their thinking is different. This made me wonder: is creative thinking really different in its very essence? So I decided as an experiment to take a watercolor class, knowing that I had never been any good at art. I wanted to get a glimpse of the emotions one goes through as a student in a subject for which one has no talent. I started the watercolor class not really understanding what techniques I should use for my brush, how to mix colors and other such technical details. But then I realized it was *only* the techniques I did not know.

I found that my aesthetic experiences with drawing in geometry gave me a feel for how to use my skill at geometric drawing in painting. Ideas of composition and perspective in painting are all so geometrical. I enjoyed reading books about composition and perspective, as well as finding out how much I already knew from my earlier geometry studies. Proportions (the golden ratio, particularly) and shapes are directly related to composition, but I had to learn about the use of colors. For perspective drawings, I already knew from three-dimensional geometric drawing how to draw in linear perspective, but I had to learn how to create an atmospheric perspective. It was crucial for me to find out that I had had similar experiences already—albeit ones obtained in different ways and for different purposes.

Below in Figure 2 is the painting I did after attending only eight watercolor classes. I started it in class and later the same day finished it at home because I could not stop. When it was dry, I looked at it and could not believe I had painted it.

Figure 2:   Daina's watercolor painting ("Sunset on Oregon Coast", 27.5" x 17.5", photograph by Daina Taimina)

Experiencing 'Undefined' Terms

In geometry, 'point' and 'straight line' are usually referred to as "undefined terms". In a formal sense, something has to be undefined, because it is impossible to define everything without being circular. However, if we want to pay attention to meanings in geometry, then we must still ask what is the meaning of 'point' and what is the meaning of 'straight'? The standard formal approach of saying these are undefined terms pushes these questions away under the carpet.

What is the meaning of 'point'?

Euclid has one answer—according to Heath's (1926/1956) translation of *The Elements*, "A point is that which has no parts" (p. 153). This is one meaning of 'point'. 'Point' has another meaning in geometry and mathematics that can be experienced by imagining zooming in on the point. A Tibetan monk/artist/geometer explained this to one of us by saying:

> Imagine a poppy seed. Now imagine in this poppy seed a temple and in the middle of the temple a Buddha and in the navel of the Buddha another poppy seed. Now in that poppy seed imagine a temple and in the temple a Buddha and in the navel of the Buddha another poppy seed. Now in that poppy seed imagine … (and keep going). Where is the point?

As we write this, we notice some similarity between this zooming and ideas in David's picture of a cat drawing a picture of a cat … .

These meanings of 'point' are not the same and, thus, bring about the following question: why and how are these meanings related? This is a *why*-question that often confronts calculus students when looking at the meanings of 'tangent', 'limit' and the 'definite integral'.

What is the meaning of 'straight'?

This is the question that starts both of the geometry books that we have written (see Henderson and Taimina, 1998, 2001a). Of course, whether a text or teacher allows this discussion or not, students (in fact, it appears, most human beings) have an experience of meanings of 'straight'. The meanings of 'straight' are part of the core foundation for meaning in geometry.

One common meaning for 'straight' is "shortest distance". This meaning can be used in practice to produce a straight line by stretching a string (or rubber band). There is another meaning in the realization that a straight line is very symmetric—for instance, "it does not turn or wiggle" or "in the plane, both sides are the same". Straight lines have at every point the following symmetries: reflection through the line, reflection perpendicular to the line, a half-turn about any point on the line, translation along the line, and so forth (see Figure 3).

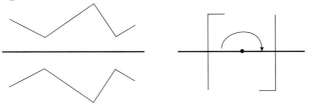

Figure 3: 'Straight' as meaning "symmetric"

This symmetry meaning is in line with Heath's (1926/1956) translation of Euclid's definition of straight line as "a line that lies evenly with the points on itself" (p. 153), which Heath then attempts to clarify in a footnote:

> we can safely say that the sort of idea which Euclid wished to express was that of a line [...] without any irregular or unsymmetrical feature distinguishing one part or side of it from another. (p. 167)

Using these experientially-based meanings of straightness, we can ask what are straight lines on the surface of a sphere. If we look at this question from a point of view outside of the sphere, then clearly the answer is that there are no straight lines on a sphere. This is the *extrinsic* point of view.

On the other hand, there is an *intrinsic* point of view. Imagine yourself to be a bug crawling on a sphere. The bug's universe is just the spherical surface. What paths on the sphere would the bug experience as straight? After some exploration, we can convince ourselves that the great circles on the sphere are the curves that have the same symmetries (with respect to the sphere) that ordinary straight lines have with respect to the plane. We thus say that the great circles are intrinsically straight. A much more usual approach in texts is simply to *define* straight lines on the sphere to be the great circles—but, again, this blocks contact with the meaning (and, thus, the potential for aesthetic experience).

So, again, why and in what way are these two meanings ("shortest" and "symmetric") related? On the sphere, we can see that (Figure 4), for two nearby points of the equator (a particular great circle), the shortest distance is along the equator. However, there is another straight path (in the sense of "symmetric") between the same two points that traverses the equator in the opposite direction (going the long way round). Thus, the "symmetric" meaning is not always the "shortest" meaning. In addition, there are surfaces with corners (see Figure 5) for which the shortest path is *not* symmetric.

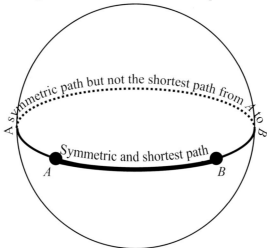

Figure 4: 'Intrinsically straight' on a sphere

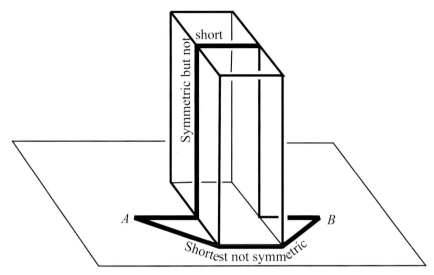

Figure 5: "Shortest" is not the same meaning as "symmetric"

A simple question that may seem intuitively straightforward at first glance, namely "what is the meaning of 'straight'?", reveals some deeper intuitions about symmetry and shortest distance, which may only become meaningful when explored in different geometrical contexts.

Proofs as Convincing Communications that Answer the Question *Why?*

Much of our own view of the nature of mathematics is intertwined with our notion of what a proof is. This is particularly true with geometry, which has traditionally been taught in high school in the context of 'two-column' proofs (see Herbst, 2002). Instead, we propose a different view of proof as "a convincing communication that answers a *why*-question".

The book entitled *Proofs Without Words* (Nelsen, 1993) contains numerous examples of visual proofs that provide an experience of *why* something is true—an experience that is, in most cases, difficult to obtain from the usual formal proofs. For example, Nelsen writes about the following result, which is usually attributed to Galileo (1615) – see Drake (1970, p. 218).

$$\frac{1 + 3 + \ldots + (2n - 1)}{(2n + 1) + (2n + 3) + \ldots + (4n - 1)} = \frac{1}{3}$$

We can easily check that this is true by simply adding the numbers:

$$\frac{1 + 3}{5 + 7} = \frac{1}{3} \qquad\qquad \frac{1 + 3 + 5}{7 + 9 + 11} = \frac{1}{3}$$

these are the cases $n = 2$ and $n = 3$ of the more general equality.

So the question is whether the general equation holds and, if so, why it holds? One way to answer the first question is to apply an argument by mathematical induction, though such an argument is unlikely to satisfy the *why*-question. Instead, look at Figure 6.

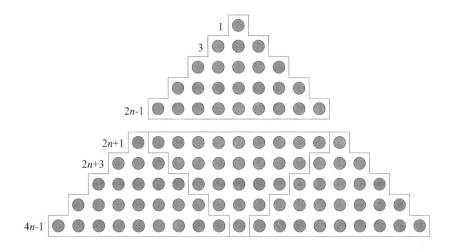

Figure 6: A proof without words (based on Nelsen, 1993, p. 115)

Through this picture, one can directly experience the meaning of Galileo's result and see both *that* is true and *why* it is true. The proof by induction would answer the question: how does Galileo's result follow from Peano's axioms? Most people (other than logicians) have little interest in that question.

> **Conclusion** In order for a proof to be an aesthetic experience for us, the proof must answer our *why*-question and relate our meanings of the concepts involved.

As further evidence toward this conclusion, many report the experience of reading a proof and following each step logically, but still not being satisfied because the proof did not lead them to experience the answers to their *why*-questions. In fact, most proofs in the literature are not written out in such a way that it is possible to follow each step in a logical, formal way. Even if they were so written, most proofs would be too long and too complicated for a person to check each step.

Furthermore, even among mathematics researchers, a formal logical proof that they can follow step-by-step is often not satisfying. For example, David's (1973) research paper ('A simplicial complex whose product with any ANR is a simplicial complex') has a very concise, simple (half-page) proof. This proof has provoked more questions from other mathematicians than any of his other research papers and most of the questions were of the sort: "Why is it true?", "Where did it come from?", "How did you see it?" They accepted the proof logically, yet were not satisfied.

Sometimes we have legitimate *why*-questions even with respect to statements traditionally accepted as axioms. One is Side-Angle-Side (or SAS):

> If two triangles have two sides and the included angle of one of them that are congruent to two sides and the included angle of the other, then the triangles themselves are congruent.

SAS is listed in some geometry textbooks as an axiom to be assumed; in others, it is listed as a theorem to be proved and in others still as a definition of the congruence of two triangles. But clearly one can ask: why is SAS true on the plane? This is especially true because SAS is false for (geodesic) triangles on the sphere. So naturally one can then ask: why is SAS true on the plane, but not on the sphere?

Here is another example – the vertical-angle theorem:

> If l and l′ are straight lines, then angle α is congruent to angle β.

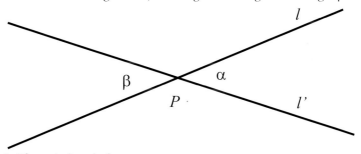

Figure 7: The vertical-angle theorem

The traditional proof of this in high-school geometry is to label the upper angle between α and β as γ, and then assert α + γ = 180° and γ + β = 180°. The usual proof then concludes that α is congruent to β because they are both equal to 180° – γ. This proof seems fine until one worries about whether the rules of arithmetic apply in this way to angles and their measures. The traditional solution in high school is to use several 'ruler and protractor' axioms to assert the properties needed. We do not know of anyone for whom this proof with the attendant axioms has aesthetic qualities (though it may be convincing). We do not usually perceive a proof as aesthetically pleasing when it is mostly repeating a list of axioms in a way that the meaning does not come through clearly. This proof seems to be an unnecessarily complicated answer to the question: why are vertical angles congruent to one another?

For about ten years of teaching this theorem in his geometry course, David was satisfied with the idea of this proof, though he managed to simplify and make more geometric the necessary assumptions contained in the 'ruler and protractor' axioms. But then one student suggested that the vertical angles were congruent because both lines had half-turn symmetry about their point of intersection, *P*. David's first reaction was that her argument could not possibly be a proof—it was too simple and did not involve everything in the standard proof. But she persisted patiently for several days and David's meanings deepened. Now her proof is much more convincing to him than the standard one, because it directly clarifies *why* the theorem is true.

Even more importantly, the meaning of the student's 'half-turn' proof is closer to the meaning in the statement of the theorem. To see this, look at the situation depicted in Figure 8.

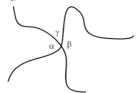

Figure 8: Are the opposite angles α *and* β *the same?*

Here, there is no symmetry: yet, the standard proof seems to apply and gives a misleading result. By means of either zooming in on the point of intersection until the curves are indistinguishable from straight-line segments (or by means of defining this angle to be the angle between the lines tangent to the curves at the intersection), symmetry arguments can be shown to apply and, hence, it is possible to argue that the angles α and β are congruent. However, the standard proof does not provide a way to discuss this, except by means of a discussion of when the 'ruler and protractor' axioms are valid.

One could ask:

> But, at least in plane geometry, isn't an angle an angle? Don't we all agree on what an angle is?

To which a reply could be:

> Well, yes and no.

Consider the acute angle depicted in Figure 9.

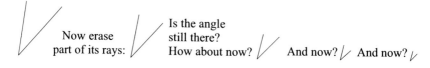

Figure 9: Where is the angle?

The angle is somehow *at the corner,* yet it is difficult to express this formally (note that the zooming meaning of 'point' seems to be involved here). As evidence of this difficulty, we have looked in all the plane geometry books in Cornell University's mathematics library for their definitions for 'angle'. We found nine different definitions. Each expressed a different meaning or aspect of 'angle' and, thus, each could potentially lead to a different proof for any theorem that crucially involves the meaning of 'angle'.

Experiencing the Hyperbolic Plane

Starting soon after Euclid's *Elements* were compiled (and continuing for the next 2000 years), mathematicians attempted either to prove Euclid's fifth postulate as a theorem (based on the other postulates) or to modify it in various ways. These attempts culminated around 1825 with Nicolai Lobachevsky and János Bolyai independently discovering a geometry that satisfies all of Euclid's postulates and common notions except that the fifth (parallel) postulate does not hold. It is this geometry that is called 'hyperbolic'. The first description of hyperbolic geometry was given in the context of Euclid's postulates and it was proved that all hyperbolic geometries are the same except for scale (in the same sense that all spheres are the same except for scale).

In the nineteenth century, mathematicians developed three so-called 'models' of hyperbolic geometry. During 1869-1871, Eugenio Beltrami and Felix Klein developed the first complete model of hyperbolic geometry (and were the first to call the specific geometry 'hyperbolic'). In the Beltrami–Klein model, the hyperbolic plane is represented by the interior of a circle, straight lines are (straight) chords of that circle and the circle's 'reflection' about a chord is a projective transformation that takes the circle to itself while still leaving the chord point-wise fixed.

Around 1880, Henri Poincaré developed two related models. In the Poincaré disc model, the hyperbolic plane is represented by the interior of a circle, with straight lines being circular arcs perpendicular to this circle. In the Poincaré upper-half-plane model, the hyperbolic plane is represented by half a plane on one side of a line, with straight lines being semi-circles that are perpendicular to this line. All three hyperbolic geometry models distort distances (in ways that are analytically describable), but the Beltrami–Klein model represents hyperbolic straight lines as Euclidean straight-line segments, while both of Poincaré's models represent angles accurately. For more details on these hyperbolic models, see Chapter 17 of Henderson and Taimina (2005a).

These models of hyperbolic geometry have a definite aesthetic appeal, especially through the great variety of repeating patterns that are possible in the hyperbolic plane. The Dutch artist M. C. Escher used patterns based on these hyperbolic models in several well-known prints (see, for example, the one in Figure 10). Repeating patterns on the sphere have an aesthetic appeal through their simplicity and finiteness. However, in these various hyperbolic models, the patterns have an aesthetic appeal for us because of their connections with infinity—there are infinitely many such patterns and each also draws us to the infinity at the edge of the disc, leaving sufficient space for our imagination.

For more than a hundred and twenty-five years, these models have been very useful for studying hyperbolic geometry mathematically. However, many students and mathematicians (including the two of us) have desired a more direct experience of hyperbolic geometry—wishing for an

Figure 10: M.C. Escher's Circle Limit III *(based on the Poincaré disc model)*
© *2004 The M.C. Escher Company*

experience similar to that of experiencing spherical geometry by means of handling a physical sphere. In other words, the experience of hyperbolic geometry available through the models did not directly include an experience of the *intrinsic* nature of hyperbolic geometry.

Mathematicians looked for surfaces that would posess the complete hyperbolic geometry, in the same sense that a sphere has the complete spherical geometry. A little earlier, in 1868, Beltrami had described a surface (called the 'pseudosphere', see Figure 11), which has hyperbolic geometry *locally*.

The pseudosphere also has a certain aesthetic appeal for us in the way (as with the Poincaré models) it points the imagination towards infinity. However, the pseudosphere allows only a very limited experience of hyperbolic geometry, because any patch on the surface that wraps around the surface or extends to the circular boundary does not have the geometry of any piece of the hyperbolic plane.

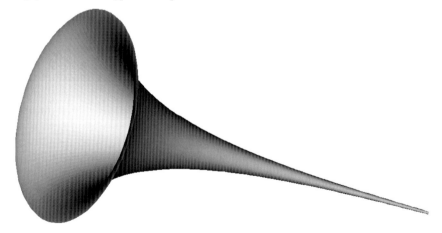

Figure 11: The pseudosphere

At the very beginning of the last century, David Hilbert (1901) proved that it is impossible to use real analytic equations to define a complete surface whose intrinsic geometry is the hyperbolic plane. In those days, 'surface' normally meant something defined by real analytic equations and so the search for a complete hyperbolic surface was abandoned. And N. V. Efimov (1964) extended Hilbert's result, by proving that there is no isometric embedding of the full hyperbolic plane into three-space, defined by functions whose first and second derivatives are continuous. Still, even today, many texts state incorrectly that a complete hyperbolic surface is impossible.

However, Nicolaas Kuiper (1955) proved the existence of complete hyperbolic surfaces defined by continuously differentiable functions, although without giving an explicit construction. Then, in the 1970s, William Thurston described the construction of a surface (one that can be made out of identical paper annuli) that closely approximates a complete hyperbolic surface. (See Figure 12 and Thurston, 1997, pp. 49-50.) The actual hyperbolic plane is obtained by letting the width of the annular strips go to zero. In 1997, Daina worked out how to crochet the hyperbolic plane, following Thurston's annular construction idea. (See Figure 13.) Directions for constructing

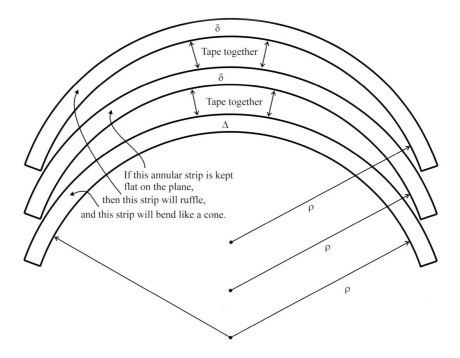

Figure 12: Construction of the annular hyperbolic plane

Thurston's surface out of paper or by crocheting can be found in Henderson and Taimina (2005a) or in Henderson and Taimina (2001b). In these refer-

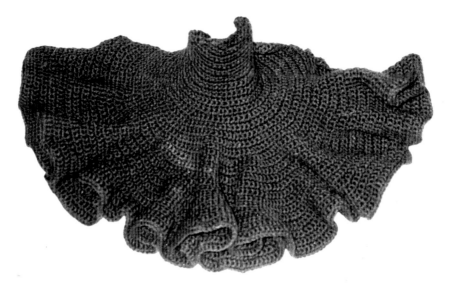

Figure 13: A crocheted hyperbolic plane (crocheted by Daina Taimina, photograph by David W. Henderson)

ences, there is also a description of an easily constructible polyhedral hyperbolic surface, called the 'hyperbolic soccer ball', comprising regular heptagons each surrounded by seven hexagons (the usual spherical soccer ball consists of regular pentagons each surrounded by five hexagons). This polyhedral surface was discovered by Keith Henderson (David's son) and provides a very accurate polyhedral approximation to the hyperbolic plane (see Figure 14).

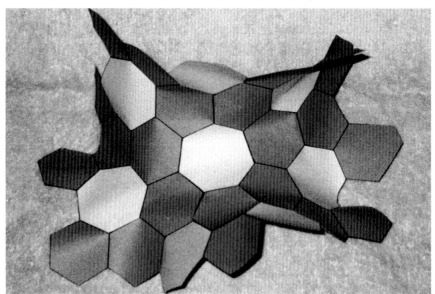

Figure 14: A hyperbolic soccer ball (made and photographed by Keith Henderson)

The geodesics ('intrinsic straight lines') on a hyperbolic surface can be found using the "symmetry" meaning of straightness discussed above: for example, the geodesics can be found by folding the surface (in the same way that folding a sheet of paper will produce a straight line on the paper). This folding also determines a reflection about that geodesic.

Now, by interacting with these surfaces, we can have a more direct experience of meanings in hyperbolic geometry. And, very importantly, we can experience the connections between these meanings and the three nineteenth-century models discussed above. These models can now be interpreted as projections (or maps) of the hyperbolic surface onto a region in the plane that distort the surface in a similar manner to the way projections (maps) of a sphere (such as the Earth) onto a region of the plane distort distances, areas and/or angles. This is important, because these models are used to study hyperbolic geometry in detail, while the surface itself allows us direct experience with the intrinsic geometry.

Before we had experience of these physical surfaces, our only experiences of hyperbolic geometry were through formal study with axiom systems and analytic study of the nineteenth-century models. The models provided aesthetic experiences that led our imagination to infinity, but this was not directly connected with geometric meanings. For example, the question that we (as well as most students) had was: why are geodesics in both Poincaré models represented by semi-circles or circular arcs?

To us, the nineteenth-century models were more like artistic abstractions. But, after constructing the surfaces, we could see how and why the geodesics are represented in the way they are. (See Henderson and Taimina, 2005a, or Henderson and Taimina, 2001a, for more details of these connections, including proofs that the intrinsic geometry of each of the surfaces is the same geometry as that represented by all of the models.)

Radius and curvature of the hyperbolic plane

Since all hyperbolic planes are the same up to scale, most treatments of the hyperbolic plane consider the curvature to be −1. It is very difficult to give meaning to the effects of the change of curvature without looking at actual physical hyperbolic surfaces with different curvatures. Each sphere has a radius r (which is extrinsic to the sphere) and its (Gaussian) curvature (as defined in differential geometry) is $1/r^2$. In a similar way, each hyperbolic plane has a radius r, which turns out to be the (*extrinsic*) radius of the annuli that go into Thurston's construction and the (Gaussian) curvature of the hyperbolic plane is $-1/r^2$. We were not aware of any meaning for the radius of a hyperbolic plane before experiencing these surfaces.

From a theoretical perspective, changing the radius or curvature is merely a change of scale and spheres, for example, of radii 4cm, 8cm and 16 cm look very much alike. However, we were shocked when we looked at the hyperbolic planes with these same radii (see Figures 15a, 15b and 15c, drawn with radii of 4 cm, 8 cm and 16 cm respectively).

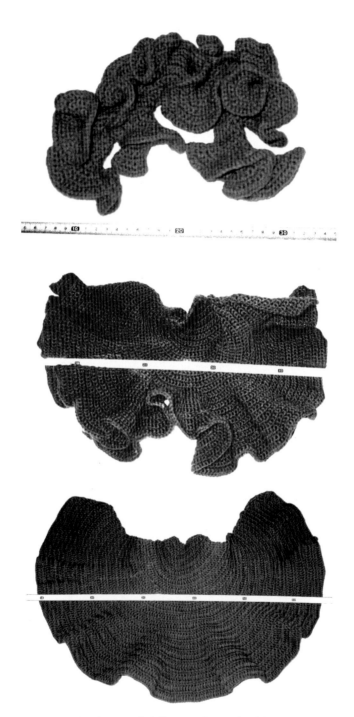

Figure 15a-c: Hyperbolic planes with different radii (crocheted by Daina Taimina, photographed by David W. Henderson)

There is a felt difference that is not present in the spheres of the same radii (the main reason for this difference seems to come from the fact of exponential growth in the hyperbolic plane). This experience of the meaning of the radius of a hyperbolic plane was a profoundly aesthetic experience for us, because we were forced to look deeper mathematically into the meanings of both radius and curvature, as well as explore the local and global natures of the hyperbolic plane.

Ideal triangles

By exploring the possible shapes of large triangles on the hyperbolic surface (see Figure 16), we can see that they seem to become more and more the same shape as they become large. This leads on to the theorem (proved by using the models) that all *ideal* triangles (namely those with vertices at infinity) are congruent and have area equal to πr^2. (This is the same as the extrinsic area of the identical circles determined by the annuli in the construction.)

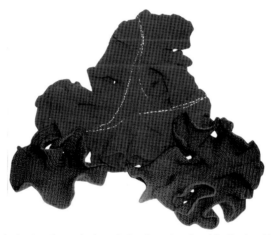

Figure 16: An ideal triangle on the hyperbolic plane (crocheted by Daina Taimina, photographed by David W. Henderson)

Horocycles or horocircles

By experiencing the annular construction (see Figure 12 once more), it is easy to see that curves perpendicular to the annuli (that is, curves that run in the radial direction) possess reflection symmetry and, thus, are geodesics. In addition, they are asymptotic to each other at infinity. Most treatments of hyperbolic geometry define *horocycles* as those curves that are orthogonal to a collection of asymptotic geodesics. Thus, the annuli (in the limit, as their width goes to zero) are horocycles. Both of us had studied hyperbolic geometry and its models; but exploring the hyperbolic surface was the first time we had experienced horocycles in a way that made clear their close connection with curvature and how, as many books simply assert, they can be described as circles with infinite (intrinsic) radii.

In the next section, we turn to look at the design of machines in the nine-teenth century—at first sight, perhaps, a surprising leap. But in a curious way, these machines embody striking geometric principles and experiences in their design and the same questions we have been addressing (such as what is 'straight'?) reappear in exciting ways and, perhaps unexpectedly, horocycles reoccur once more.

Experiencing Geometry in Machines

Recently, we have been working on an NSF-funded project to examine the mathematics inherent in a collection of nineteenth-century mechanisms, as well as to see to the inclusion of these mechanisms (along with commen-taries and learning modules) as part of the new National Science Digital Library (NSDL—see www.nsdl.org). Our experiences with these all of vari-ous mechanisms are offering us different perspectives on geometry, per-spectives that arise from motion. For example, this work has brought us back to the question: what is 'straight'?

When using a compass to draw a circle, we are not starting with a figure we accept as circular: instead, we are using a fundamental property of circles, namely that the points on a circle are at a fixed distance from the center, as the basis for the tool. In other words, we are drawing on a mathematical definition of a circle. Is there a comparable tool (serving the equivalent role to a compass) that will draw a straight line? If, in this case, we want to use Euclid's definition ("a straight line is a line that lies evenly with the points on itself"), this will not be of much help.

One could say:

> We use a straight-edge for constructing a straight line.

To which a response might be:

> Well, how do you know that your straight-edge is straight? How do you know that anything is straight? How can you check that some-thing is straight?

This question was important for James Watt. When he was thinking about improving steam engines, he needed a mechanism in order to convert cir-cular motion into straight-line motion and *vice versa*. In 1784, Watt found a practical solution (which he called "parallel motion") that consisted of a link-age with six links. He described his parallel motion mechanism as being free of "untowardly frictions and other pieces of clumsiness", claiming it to be "one of the most ingenious simple pieces of mechanisms that I have con-trived" (in Ferguson, 1962, p. 195). These expressions of smoothness and efficiency seem to be very close to what we are calling 'aesthetic'. However, Watt's mechanism produced only approximate straight-line motion: in fact, it actually produces a stretched-out figure of eight. Mathematicians were not satisfied with this approximate solution and worked for almost a hundred years to find exact solutions to the problem. A linkage that draws an exact straight line (see Figure 17a) was first reported by Peaucellier, in 1864. (See Henderson and Taimina, 2005a, 2005b, for a discussion of relevant history.)

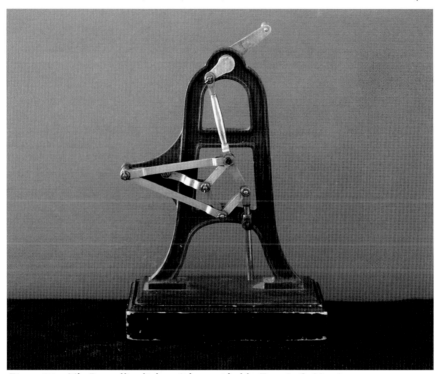

Figure 17a: The Peaucellier linkage (photographed by Francis C. Moon)

Why does the Peaucellier linkage draw a straight line? We suggest the reader connect to a web site where this linkage is depicted in motion (for example, see: KMODDL.library.cornell.edu). As an exercise in analytic geometry, one can verify *that* the point Q will always lie along a straight line—but this still does not answer the *why*-question. Especially difficult is being able to see any relationship with either the "shortest" or the "symmetric" meaning of straightness: is there perhaps a different meaning of straightness that is operative here?

In the 'inversor' (that is, the links joining C, R, Q, S, and P in Figure 17b), the points P and Q are inverse pairs with respect to a circle with center C and radius $r = \sqrt{s^2 - d^2}$. Analytically, this means that:

distance (C to P) × distance (C to Q) = r^2.

Here, the crucial property of circle inversion is that it takes circles to circles. (For details on circle inversion, see Chapter 16 of Henderson and Taimina, 2005a.) After experiencing the motion of the linkage, we now see that P is constrained (by means of its link to the stationary point B) to travel in a circle around B. Thus, Q must be traveling along the arc of a circle. The radius

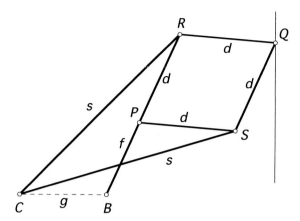

Figure 17b: The Peaucellier linkage diagram

and center of this circle can be varied by changing the position of the fixed point B and the length of the link BP.

Thus, the Peaucellier linkage draws (at Q) the arc of a circle without reference to the center of that circle. If the lengths of CB and BP are equal, then the circle on which P moves goes through the center C. Since points near C are inverted to points near infinity, the circle that Q lies on must go through infinity. How can a circle go through infinity? Answer: only if the circle has infinite radius. *A circle with infinite radius (and thus zero curvature) is a straight line.* We now have a third meaning for straight line—and the Peaucellier linkage is a tool for drawing a straight line that draws on this meaning.

In the previous section on hyperbolic geometry, we pointed out that the *horocycles* in the hyperbolic plane can be seen as circles of infinite radius. Thus, circles with infinite radius are not straight in the hyperbolic plane, even though they are straight in the Euclidean plane. This proves that seeing 'straight' as "circle of infinite radius" is a different meaning from either 'straight' as meaning "symmetric" or 'straight' as meaning "shortest".

Behind this discussion lies the theory of circle inversions, one of the most aesthetic geometric transformations that have also been used in modern art. The special aesthetic appeal here is that inversions (as seen in Figure 18) can draw out the imagination to infinity and can also bring out important geometric meanings. For example, the experience of the linkage as a mechanism that draws a circle without using its center allows one to understand how the linkage can draw a circle of infinite radius and, thus, a straight line.

Peaucellier's linkage is one of thirty-nine straight-line mechanisms in Cornell University's collection, which also has more then two hundred and twenty kinematic models designed by Franz Reuleaux. These models are a rediscovery of a lost, nineteenth-century machine design knowledge. Franz Reuleaux is often referred to as a 'father of modern machine design' (see, for example, Moon, 2003, p. 261). Reuleaux's two most important books

Figure 18: An example of inversion-based art (M.C. Escher's Development II, *1939*
© *2004 The M.C. Escher Company)*

contain hundreds of drawings of machines and mechanisms. To complement his books, Reuleaux designed and built over eight hundred kinematic
models to illustrate his theory of machines. The models in the Cornell collection clearly show the aesthetic style of Reuleaux. (To read more about
Reuleaux, his mechanisms and his theory of machines, see Moon, 2003,
which also contains many further references.)

As we have been exploring the mathematics behind the Reuleaux models
for the NSDL project, we are repeatedly surprised how much aesthetic
appeal we find there—not only in machine design itself, but also in the
mathematics. These experiences caused us to ask about the relationships
among mathematics, engineering and art. Leonardo da Vinci is a well-
known embodiment of this interrelationship, but we have found that there
seems to be a broader connection. For example, Reuleaux, in his book *The
Kinematics of Machinery* (1876/1963), refers specifically to the artist and to
experiences of deeper meanings in a manner similar to our discussion at the
beginning of this chapter.

> He who best understands the machine, who is best acquainted
> with its essential nature, will be able to accomplish the most by its
> means. (p. 2)

> In each new region of intellectual creation the inventor works as
> does the artist. His genius steps lightly over the airy masonry of
> reasoning which it has thrown across to the new standpoint. It is
> useless to demand from either artist or inventor an account of his
> steps. (p. 6)

> The real cause of the insufficiency of [previous classification sys-
> tems] is not, however, the classification itself; it must be looked for
> deeper. It lies [...] in the circumstance that the investigations have
> never been carried back far enough, – back to the rise of the ideas;
> that classification has been attempted without any real comprehen-
> sion being obtained of the subjects to be classified. (p. 18)

In addition, in his article on the history of engineering, Eugene Ferguson
(1992) wrote:

> Both the engineer and the artist start with a blank page. Each will
> transfer to it the vision in his mind's eye. The choice made by
> artists as they construct their pictures may appear to be quite arbi-
> trary, but those choices are guided by the goal of transmitting their
> visions, complete with insights and meaning, to other minds. [...]
> The engineers' goal of producing a drawing of a device—a
> machine or structure or system—may seem to rule out most if not
> all arbitrary choices. Yet engineering design is surprisingly open-
> ended. A goal may be reached by many, many different paths,
> some of which are better than others but none of which is in all
> respects the one best way. (p. 23)

Ferguson also notes that Robert Fulton (of steamboat fame) and Samuel
Morse (the inventor of the electrical telegraph) were both professional artists
before they turned to careers in technology.

We have already mentioned the Peaucellier linkage. Another example is
Reuleaux triangles, which are the most well-known of curves with constant
width. If a closed convex curve is placed between two parallel lines and the
lines are moved together until they touch the curve, the distance between
the parallel lines is the curve's 'width' in one direction. Because a circle has
the same width in all directions, it can be rotated between two parallel lines
without altering the distance between the lines.

The simplest, non-circular, constant-width curve is known as the
Reuleaux triangle. Mathematicians knew it earlier, but Reuleaux (1876/
1963, pp. 131-146) was the first to study various motions determined by constant-width figures. A Reuleaux triangle can be constructed starting with an
equilateral triangle of side s and then replacing each side by a circular arc
using the other two sides as radii, as shown in Figure 19. The resulting figure bounded by these three arcs is the Reuleaux triangle. Its constant width
is equal to s, the side length of the original equilateral triangle.

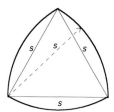

Figure 19: A Reuleaux triangle

In Reuleaux's collection, we find several applications of this triangle and other constant-width curves: see, for example, Figure 20.

Figure 20: A Reuleaux mechanism using a constant-width triangle
(photographed by Francis C. Moon)

The Reuleaux triangle fits inside a square of side s and can be rotated a full 360° within the square—this is the idea behind drill bits that can drill (almost) a square hole: conversely, the square can rotate around the stationary Reuleaux triangle. Reuleaux did not give analytical descriptions of these motions. Instead, he produced many drawings that, in an aesthetically visual way, show the different paths of points during the motions.

Reuleaux was the first to describe properties of these motions accurately and, in his model collection, we find several applications, such as those illustrated above. For instance, he proved the following theorem geometrically: any relative motion between two shapes, S and R, in the plane can be realized as the motion of two other shapes, cS and cR, rolling on each other, with cS fixed to S and cR attached to R. He called the rolling shapes 'centroids' (the locus of instantaneous centers), but, in order to avoid confusion with the centroid of a triangle, the word 'centrode' was subsequently used.

Figures 21 and 22 (overleaf) show the centrodes (namely $O_1O_2O_3O_4$ and $m_1m_2m_3$) for the relative motions of the square and the Reuleaux triangle respectively. Since the *relative* motions are the same in the two figures, the centrodes are necessarily the same. But the real meaning of this rolling motion can be experienced only by actually looking at the models in motion.

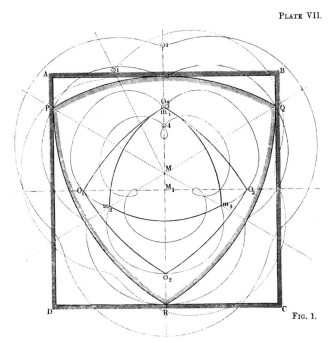

Figure 21: A Reuleaux triangle moving in a square (from Reuleaux, 1876/1963, p. 136)

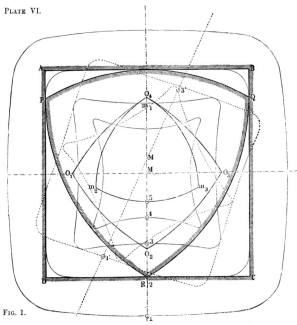

Figure 22: A square moving around a Reuleaux triangle (from Reuleaux, 1876/1963, p. 137)

Conclusion

Aesthetics has always been a driving force in our experiences of mathematics. We do not—as some mathematicians have claimed to do—carry with us a list of criteria by which we judge the aesthetic value of a proof. In fact, rarely do we find proofs, in and of themselves, to be aesthetic objects. Instead, we locate the aesthetic value of mathematics in the coming-to-understanding, in the *integration* of experience and meaning. We believe that the understanding of meanings in mathematics (often through aesthetic experiences) comes *before* an understanding of the analytic formalisms. We hope that the reader has gained, through our stories and our examples, a sense of the aesthetic component of our perception of mathematical meanings.

Acknowledgement

Partial support for this work was provided by the US National Science Foundation's National Science, Technology, Engineering, and Mathematics Education Digital Library (NSDL) program, under grant DUE-0226238.

Section B

A Sense for Mathematics

Introduction to Section B

The three chapters in Section B, *A Sense for Mathematics,* offer perspectives from authors with backgrounds in mathematics and philosophy who strive to elucidate and explain the aesthetic dimension of mathematics. The focus of these chapters is pan-mathematical, crossing specific fields and historical periods. The authors explore aesthetic issues related to mathematical understanding, the development of mathematical knowledge and public perceptions of mathematics and mathematicians.

In Chapter 4, Nathalie Sinclair proposes a model of the aesthetic nature of mathematical inquiry. Although mathematicians tend to privilege the beauty and elegance of their proofs, theorems and other mathematical entities – their works of art – a closer examination of their processes of inquiry reveals important, perhaps necessary, roles played by the aesthetic in the discovery and development of mathematical knowledge. Based on her interviews with professional mathematicians, as well as reports found in the literature, Sinclair identifies and illustrates three distinct roles of the aesthetic. These mathematicians' aesthetic claims relating to these roles are then analysed and explained in terms of contemporary theories of inquiry and experience.

Martin Schiralli, in Chapter 5, attempts an analysis of the key concept of 'pattern' and examines the way 'pattern' functions in the thinking of mathematicians about mathematics through the ages. He is especially interested in the way the aesthetic insinuates itself – seemingly ineluctably – into that thinking. By focusing on certain notions of *arithmos* (number) most prevalent and powerful in Pythagorean thinking, Schiralli links some of the earliest aspects of mathematical–aesthetic thinking with influential contemporary views on mathematics as the 'science' of patterns. More broadly, the pervasive use of the idea of pattern in Gregory Bateson's work in biology and 'the ecology of mind', as well as in the work of art historian Ernst Gombrich on 'the sense of order', inform this chapter. Both Bateson (on science and epistemology) and Gombrich (on abstract and decorative art) offer perspectives on the correspondence between the aesthetic and the mathematical that may profitably be linked back to these early Greek views.

In Chapter 6, William Higginson situates the book's investigation on the connection between mathematics and the aesthetic in a much broader intellectual terrain, namely around the age-old question of 'What does it mean to be human?' He both enriches and complicates the connection on the one hand, by probing perspectives on mathematics and mathematicians in the broader culture, including social scientists, schoolchildren, playwrights, moviemakers and novelists, and on the other, by drawing on researchers' insights in cognitive science, philosophy and anthropology regarding the aesthetic nature of human activity. He argues that the roots of mathematical activity are located perhaps surprisingly close to this human aesthetic predisposition.

CHAPTER 4
The Aesthetic Sensibilities of Mathematicians

Nathalie Sinclair

I begin with a story told by the French mathematician François Le Lionnais (1983) about his first experience, at age seven, of a mathematical discovery. It illustrates, perhaps more immediately than a trip to the Great Museum of 'elegant' mathematical proofs, how aesthetic responses, values and experiences can snugly insinuate themselves alongside logical steps and decisions throughout mathematical activity.

The young Le Lionnais is sitting alone at the family kitchen table, with a pencil and some paper, not tired enough to join the rest of his family for the afternoon siesta. Seated at the table, he writes the numbers from 1 to 9. But instead of multiplying one number by the others, as he has been learning to do in school, he multiplies each by itself, before writing the result in the row beneath.

1	2	3	4	5	6	7	8	9
1	4	9	16	25	36	49	64	81

"Suddenly", he writes, describing his memory of the experience, "a veil lifts, allowing me to perceive in this otherwise dull alignment a beautiful structure" (p. 12; *my translation, as are all quotations from his account*). But to see it, he continues, one has to consent to "an amputation" (p. 12), striking out the digits in the tens decimal place, conserving only the units. This produces:

1	2	3	4	5	6	7	8	9
1	4	9	6	5	6	9	4	1

Le Lionnais admits that an ordinary adult might have found the resulting symmetry (with respect to the middle number) quite banal, but, having discovered it himself, he is thunderstruck. He feels he has entered a "vast domain where a multitude of treasures has been hidden" (p. 13). Surely, he could continue mining his beautiful array simply by multiplying the digits once again (and maybe even again), remembering to strike out the tens decimal place.

1	2	3	4	5	6	7	8	9
1	4	9	6	5	6	9	4	1
1	8	7	4	5	6	3	2	9

Hélas! The symmetry has been lost. Le Lionnais is stubborn, though, certain that "chaos could not have taken over the society of numbers, which had thus far been so well organised" (p. 13). And all of a sudden, he sees it: the digits occupying the symmetric positions are complements of each other in ten: $1 + 9 = 10$; $8 + 2 = 10$; $7 + 3 = 10$; $4 + 6 = 10$. Now he can move on: what happens in the fourth row? Le Lionnais works on his rows of digits all afternoon until he is called for dinner, finding that the sixth row reproduces the second, while the seventh reproduces the third, and so on.

Le Lionnais recounts this story in a prelude to his quixotic book of numbers, as a way of describing how he, himself, became bewitched by that queenly domain of mathematics, the theory of numbers. Remarkably, instead of extolling the great beauty of Euclid's proof of the infinity of primes or the Pythagorean proof of the irrationality of $\sqrt{2}$ – as so many other mathematicians have done – Le Lionnais describes the beauty he sees in the process of playing with, manipulating and transforming simple whole numbers.

He has not produced a proof, or even a theorem – the traditional products with aesthetic currency in mathematics. Rather, he has sought out (and found) pleasing and generative instances of symmetry, balance and pattern in the magical, self-sufficient world of numbers. When he lifts the veil and notes the symmetry in his row of numbers, he perceives the rich potential of his simple rule (multiply each number by itself) that motivates him to explore further; after all, as he remarks, symmetry does not happen by accident.

The evaluative role of the aesthetic – in which properties such as 'beauty' and 'elegance' are used to distinguish good from not-so-good mathematical products – has been quite well documented by mathematicians. As was mentioned in Chapter α, a number of mathematicians, such as Hardy (1940), have even offered lists of aesthetic criteria that can be applied to determine a product's aesthetic value, perhaps because they felt that a proof's 'beauty' and 'elegance' should be as timeless and objective as its truth. However, Le Lionnais's story evokes two additional roles for the aesthetic over and above the purely evaluative. On the one hand, he is guided by his response to qualities such as symmetry and balance; on the other, he is motivated by the unexpected treasures he finds as he plays with numbers. The aesthetic both *organises* and *motivates* his mathematical activity.

In the following sections, I flesh out and broaden these three roles of the aesthetic – what I call the evaluative, the generative and the motivational – each of which turns out to be far more pervasive in and fundamental to the development of mathematics than might be suggested by a single story of a precocious boy playing with numbers. There are several thorny issues to contend with, including the very definitions of the words 'aesthetic', 'beauty' and 'elegance'. While different mathematicians may retain personal, idiosyncratic meanings for 'beauty' and 'elegance', I will be using the word 'aesthetic' in a quite specific sense, namely that of the American philosopher John Dewey (1934). Dewey claimed experiences, responses and objects have an aesthetic quality when they provoke a pleasurable 'sense of fit' for the individual.

Thus, the aesthetic, for Dewey, pertains to decisions about pleasure as well as meaning, thereby operating on both affective and cognitive levels. Objects do not, in and of themselves, possess aesthetic qualities [1]; they require a perceiver as well as a socio-historical context. Cultural differences influence aesthetic responses: yet, given the large degree of communication now possible, the differences in responses between contemporary Greek and ancient Greek mathematicians are certainly greater than those between, say, twenty-first-century North American and Japanese mathematicians. However, since I am presently concerned with the way the aesthetic deploys itself across the spectrum of the contemporary mathematics milieu, I forego a socio-historical analysis.

In the rest of this chapter, I propose a tripartite categorisation for structuring the diversity of aesthetic responses found in the mathematics literature.

(a) The most obvious and public of the three characteristics is the *evaluative*; it concerns the aesthetic value of mathematical products such as results or proofs and, more specifically, the judgements made about which products are most significant. Mathematicians may evaluate both their own work, as they complete a proof or solution, as well as the work of others, as they review potential journal articles or attend colloquia.

(b) The *generative* characteristic of the aesthetic pertains to those aesthetic modes of reasoning used in solving problems, as opposed to logical or even intuitive ones. I have used the term 'generative', because it is described by mathematicians as being responsible for generating new ideas and insights that could not be derived by logical steps alone (see, for example, Poincaré, 1908/1956).

(c) Lastly, the *motivational* characteristic relates to the role of the aesthetic in attracting mathematicians to certain fields and, in turn, in stimulating them to work on certain problems. While the evaluative characteristic of aesthetics operates on mathematicians' finished, public work, the motivational and generative characteristics belong to more private, evolving facets of mathematical inquiry.

The Evaluative Characteristic of the Aesthetic

Of the hundreds of thousands of theorems that are now proved each year, how do mathematicians select which theorems become a part of the body of mathematical knowledge? Which are to be those that get printed in journals or presented at conferences, which are deemed worthy of being further developed, as well as worth being taught to students or placed in textbooks?

Tymoczko (1993) argues that aesthetic criteria are necessary in order to ground value judgements in mathematics (judgements such as importance and relevance) for two reasons. First, selection is essential in a world where infinitely many correct theorems could be produced. Second, mathematical reality cannot provide its own criteria: that is, a mathematical result can not be judged important because it matches some supposed mathematical reality – mathematics is not self-organised. In fact, it is only in relation to actual mathematicians with actual interests and values that mathematical reality is divided up between the trivial and the important.

Steiner (1998) goes even further in claiming that there is no objective criterion for a certain structure to qualify as mathematics – and not every structure counts as a mathematical structure (chess, for example). Instead, he claims mathematicians decide what structures are 'mathematical' based on the aesthetic criteria of beauty and convenience. Both beauty and convenience are anthropocentric (or 'species-specific') notions, because they are based on the needs and limitations of *human beings*. If human brains were a thousand times more powerful, they might have no need for a convenient concept such as the logarithm.

Hardy (1940) privileged supposedly objective criteria such as depth and generality, but his more 'purely aesthetic' criteria such as unexpectedness, inevitability and economy certainly play a role in determining the value of mathematical products. For example, most mathematicians agree that the Riemann hypothesis is a significant problem – perhaps because it is so intertwined with other results or perhaps because it is somewhat surprising – but its solution (if and when it comes) will not necessarily itself be considered 'beautiful'. That judgement will depend on many things, including the knowledge and experience of the mathematician in question and whether it illuminates any of the many connections mathematicians have identified or whether it renders them too obvious. [2]

Buckminster Fuller (quoted in Fadiman, 1985) said that, when working on a problem, he "never thinks about beauty" (p. 85). It is only if he finds that "the solution is not beautiful" (p. 85) that he knows it is wrong. His statement, echoed by many others, is misleading: do mathematicians only think about the beauty of solutions after the event? The work of Le Lionnais (1948/1986) illustrates many other mathematical entities, other than proofs and solutions, which are amenable to aesthetic consideration, including images, definitions, methods of proving and concepts themselves. He also treated the mathematical aesthetic in terms of a matrix of two principles: the structure of mathematicians' works and a human's conception of beauty. Following Nietzsche, Le Lionnais saw an individual's conception of beauty as falling into two categories – classicism and romanticism – which parallel the Apollonian and Dionysian ones. These categories represent two styles of human endeavour: on the one hand, a desire for equilibrium, harmony and order; on the other, a yearning for lack of balance, form obliteration and pathology. [3]

For Le Lionnais, classically beautiful 'facts' are ones whose beauty impresses through austerity or mastery over diversity, such as magic squares and Pascal's triangle. [4] Romantically beautiful 'facts', such as the imaginary numbers, impress through "le culte des émotions violentes, du non-conformisme et de la bizarrerie" (p. 444). These engaged, even ecstatic, descriptions of romantically appealing mathematical facts imply the existence of intemperate aesthetic responses antithetical to ones such as detachment which are usually articulated with such equanimity – they speak of a radical, emotional, individualised component to aesthetic response. Classically beautiful 'methods' permit the attainment of powerful effects through moderation, such as proof by recurrence or the notion of the locus. On the other hand, romantically beautiful 'methods', such as *reductio ad absurdum* or non-constructive existence proofs, are characterised by indirectness: failing to shed light on the mathematical structure, they leave one in a state of conflict or even that of dissatisfaction.

That mathematicians can respond aesthetically to the wide range of their tools and materials, and not only to their solutions and proofs which involve a truth component, makes it much easier to see how different mathematicians come to choose their specific research domains. In addition, the mathematician's constant interaction with mathematical tools and materials, in the course of inquiry, explains how aesthetic judgements can easily affect the *process* of inquiry and not just its final product. For example, in an interview with me [5], Jonathan Borwein observed, "I would emphasise how many mathematicians will abandon a proof technique they are 'sure' will work, because it would be dull, ugly, inelegant". However, before pursuing this idea, I want to continue focusing on mathematicians' aesthetic evaluation of mathematical products. But this time, however, I want to attend more closely to the way they are shared within the mathematical community.

The aesthetic dimension of mathematical expression and communication

The evaluative characteristic of the aesthetic is not only involved in judging the great theorems of the past or existing mathematical entities, but it is also actively involved in mathematicians' decisions about expressing and communicating their own work. As Wolfgang Krull (1930/1987) wrote:

> Mathematicians are not concerned merely with finding and proving theorems, they also want to arrange and assemble the theorems so that they appear not only correct but evident and compelling. (p. 49)

In the same interview with me as mentioned above, Borwein noted that some mathematicians derive "the most pleasure in refining, polishing and harvesting their conquests at this stage". This would certainly seem to be the case for Gauss, who presented no less than six different proofs of the law of quadratic reciprocity in his *Disquisitiones Arithmeticae* (1801/1966). In an article published in 1817, discussing his various proofs of this result, Gauss wrote about his own quest for beauty and simplicity, defending it from charges of redundancy.

> As soon as a new result is discovered by induction, one must consider as the first requirement the finding of a proof by *any possible* means. But after such good fortune, one must not in higher arithmetic consider the investigation closed or view the search for other

> proofs as a superfluous luxury. For sometimes one does not at first
> come upon the most beautiful and simplest proof, and then it is
> just the insight into the wonderful concatenation of truth in higher
> arithmetic that is the chief attraction for study and often leads to the
> discovery of new truths. For these reasons the finding of new proofs
> for known truths is often at least as important as the discovery itself.
> (Gauss, 1863, pp. 159-160; in May, 1972, p. 299)

Several aesthetic qualities I identified in the previous sub-section are operative at this stage of the mathematician's inquiry as well. For instance, while
some mathematicians may provide the genesis of a result, as well as logical
and intuitive substantiation, others prefer a 'clean', 'pure' or 'minimal' presentation of *only* the logically formed results, of *only* the elements needed to
reveal the structure. On the other hand, Philip Davis (1997) wants his proofs
to be transparent:

> I wanted to append to the figure a few lines, so ingeniously placed
> that the whole matter would be exposed to the naked eye. I wanted
> to be able to say not ὅπερ ἔδει δεῖξαι *(quod erat demonstrandum)*,
> as did the ancient Greek mathematicians, but simply, "Lo and
> behold! The matter is as plain as the nose on your face." (p. 17)

Thus, the mathematician's aesthetic judgements also affect the way she
organises her exposition, whether opting for an intuitive, perhaps visually-
oriented proof, a detailed proof with examples or a short, abstract proof. In
these cases, as well as those in between, the mathematician is expressing an
understanding according to a personal aesthetic. Of course, there is a strong
enculturation that takes place. In interview, Robert Osserman claimed that
younger mathematicians will use this last method of presentation as a
default. They may even accept this method as the way things are or even
should be (see Chapter 8 for more on this). In his conversation with me, Joe
Buhler noted, "It took me a long time to realise that some of my most
admired role models were crummy expositors!"

However, Osserman went on to point out that it is not uncommon to
find more seasoned and established mathematicians communicating in narratives, by including in their expositions arguments of relevance, connection
and personal interest. Some mathematicians bemoan the dry, terse form of
most proofs that can often obscure the motivations and paths that led to
them (see Burton, 2004). These mathematicians are interested in the motivations behind results, the false starts and lucky guesses that led to the
results and the possibility of 'seeing' what is beautiful and interesting about
the result. In fact, behind the terse face of journal articles and textbooks,
there is a world of aesthetic persuasion. Buhler talked about having to 'mar-
ket' ideas, adding "you have to take whatever turned you on about it and
try to communicate that to someone". This is not about convincing colleagues or readers of truth or correctness; it is about convincing them about
interest and attractiveness, about how it 'connects' to them.

These judgements of beauty, elegance and worth are not frivolous ones: rather, they contribute to the on-going negotiation concerning which are the problems worth attending to, worth solving, which solutions are acceptable and what contributes to greater mathematical understanding. When Tymoczko (1993) advocates greater explicit aesthetic criticism in mathematics, he is not implying that aesthetic judgements are non-existent in the mathematics community. On the contrary, they are pervasive and operative, as suggested in the above descriptions, yet are often implicit and rarely made public.

In addition, the mathematicians quoted above highlight the role of the evaluative characteristic of the aesthetic in their own formulation and presentation of results, thus revealing a very personal side to aesthetic judgement of mathematical products. They describe the aesthetic decisions involved in expressing their results in the most satisfying ways, in much the same way as poets describe the aesthetics of expressing their various thoughts in certain forms. [6] Thus, the evaluative characteristic of the aesthetic is operational not only in the community's decisions about the significance of results, but also in the mathematician's individual decisions about the personal value of those results.

The Generative Characteristic of the Aesthetic

The generative characteristic of the aesthetic may be the most difficult of the three to discuss explicitly, operating as it most often does at a tacit or even sub-conscious level and intertwined as it frequently is with intuitive modes of thinking. The generative characteristic of the aesthetic is involved in the actual process of inquiry, in the discovery and the invention of solutions or ideas. It guides the actions and choices that mathematicians make as they try to make sense of objects and relations.

Some background

Henri Poincaré (1908/1956) was one of the first modern mathematicians to draw attention to the aesthetic dimension of mathematical invention and creation. According to his account, two operations take place in mathematical invention: first the *construction* of possible combinations of ideas and then the *selection* of the fruitful ones. Thus, to invent is to choose useful combinations from the numerous ones available, the useful ones being those that are the most beautiful, those best able to "charm this special sensibility that all mathematicians know" (p. 2048).

Poincaré explained that such combinations of ideas are quite harmoniously disposed, so the mind can effortlessly embrace their totality without realising their details. It is this harmony that at once satisfies the mind's aesthetic sensibilities and acts as an aid to the mind, sustaining and guiding it. He claimed that the sorting of combinations of ideas must happen in the unconscious, since mathematicians only become aware of the ones that already have the stamp of beauty.

This may sound a bit far-fetched, but there seems to be some scientific basis for it. The contemporary neuroscientist Antonio Damasio (1994) points out that because humans are not parallel processors, they must somehow filter the multitude of stimuli coming in from the environment. Some kind of pre-selection is carried out, whether covertly or not.

A concrete example might help to illustrate Poincaré's claims. Silver and Metzger (1989) report on a mathematician's attempts to solve a number-theory problem. (Prove that there are no prime numbers in the infinite sequence of integers 10001, 100010001, 1000100010001, ...) In working through the problem, the subject hits upon a certain prime factorisation, namely 137 x 73, that he describes as being "wonderful with those patterns" (p. 67). Something about the surface symmetry of these factors appeals to the mathematician, leading him to believe that they might go down a generative path. (The young Le Lionnais had a very similar generative experience when he perceived the symmetry in his second row of numbers and yet another when he discerned the complementary balance inherent in the third row of numbers.)

Based on their observations, Silver and Metzger also argue that aesthetic monitoring is not strictly cognitive, but appears to have a strong affective component:

> decisions or evaluations based on aesthetic considerations are often made because the problem solver 'feels' he or she should do so because he or she is satisfied or dissatisfied with a method or result. (p. 70)

The above example illustrates how an aesthetic response to a certain configuration is generative, in that it serves to lead the mathematician down a certain path of inquiry. This path is not chosen for logical reasons but, rather, because the mathematician *feels* that the appealing configuration should reveal some insight or fact.

There is a range of stimuli that can trigger aesthetic responses: a quality such as symmetry might do so, but more subtle qualities such as the 'prettiness' of an equation or the sudden emergence of a new quantity can also act as triggers. This example also illustrates how mathematicians must believe in, and trust, their feelings in order to exploit the generative characteristic of the aesthetic. They must view mathematics as a domain of inquiry where phenomena such as feeling and intuition play an important role alongside hard work and logical reasoning.

Evoking the generative characteristic of the aesthetic

There are also some special strategies that mathematicians use during the course of inquiry which seem to be oriented toward triggering the generative characteristic of the aesthetic. I will discuss four such strategies:

- playing with or 'getting a feel for' a situation;
- establishing intimacy;

- enjoying the craft;
- capitalising on intuition.

The phase of playing around or 'getting a feel for' is aesthetic in so far as the mathematician is framing an area of exploration, qualitatively trying to fit things together and seeking patterns that connect or integrate. Helen Featherstone (2000) terms this 'mathematical play', drawing on Johan Huizinga's (1950) theory of play. Huizinga saw play as the free, orderly and aesthetic exploration of a situation. The exploration is aesthetic in that the one playing is seeking to identify organising themes and structures and to arrange the objects being played with in a meaningful, expressive way.

Play is neither random, nor does it have the ultimate goal that solving problems has: rather, its goal *is* the exploration itself. In seeing play in this way, Huizinga called attention to the possibility that, in 'mathematical play', the mathematician is aesthetically exploring a certain terrain, trying to impose structures and generate patterns. And, in the course of such play, structures and patterns are indeed revealed.

Secondly, mathematicians seem to develop a personal, intimate relationship with the objects they work with, as can be evidenced by the way they anthropomorphise them or coin special names in an attempt to hold them, to own them. For example, Douglas Hofstadter (1992) first baptises his emerging object "my Magic Triangle" and then "my hemiolic crystal" (pp. 9-11). Paul Lévy (1970) becomes equally possessive about the objects; he insists on referring to the focus of his investigations as *"ma courbe"* (p. 20), even though it is generally known as the von Koch curve. Possessively naming these objects makes them easier to refer to and may even foreshadow identifying their properties. Equally as important, though, it gives the mathematician some traction on the still-vague territory, some way of marking what she *does* understand.

Norbert Wiener (1956) did not underestimate these attempts to operate with vague ideas. He recognised the mathematician's:

> power to operate with temporary emotional symbols and to organize out of them a semi-permanent, recallable language. If one is not able to do this, one is likely to find that his ideas evaporate from the sheer difficulty of preserving them in an as-yet unformulated shape. (p. 86)

Verena Huber-Dyson (1998) also evokes this unformulated, tacit knowledge:

> All the while you are aware of the pattern [...], just below the threshold of consciousness, exactly as a driver is aware of the traffic laws and of the coordinated efforts of his body and his jeep. That is how you find your way through the maze of mathematical possibilities to the 'interesting' [cases]. (p. 2)

Thirdly, there is a certain amount of craft in the mathematician's work that is also aesthetic in nature. Mathematicians have tools that can be used to create new mathematical objects or transform existing ones. Le Lionnais's (1948/1986) various kinds of 'methods' used to work with mathematical entities can also be thought of as tools.

Adrian Lewis (in his interview with me) described the use of certain tools as part of the aesthetic dimension of enjoying his craft, of using "well-worn tools in often routine ways, like a well-oiled piece of engineering". What he finds beautiful is "not just the startling revelation or the philosophical wonder" of a work of mathematics, but the *craft* of it, "the inexorable sequence of simple tools at work". Although less dramatic than a startling revelation, for him there is something comforting in the knowledge that the careful application of a tool will produce a "perfect, fine-tuned result".

In comparing science and art, Freeman Dyson also commented on:

> [the] aesthetic pleasure of the craftsmanship of performance [particularly in mathematics]. And if one is handling mathematical tools with some sophistication it is a very nonverbal and a very, very pleasurable experience just to know how to handle the tools well. It's a great joy. (Dyson *et al.,* 1982, p. 139)

Doris Schattschneider (in Chapter 2) provides additional insight into this notion of craft, when she describes some of the 'paradigms' that mathematicians use in the course of solving problems and proving theorems. For example, a symmetry argument (or the pigeon-hole principle) is used as a way to transform an unknown complex situation into a simpler, more familiar one. It may also provide insight into the structure of the unknown situation. Schattschneider views these paradigms as beautiful, because of their powerful ability to simplify, to cut across complexity and surface differences or to reformulate a problem in more familiar terms. Also, these paradigms may still carry vestiges of the aesthetic impact they had when mathematicians first encountered them in a proof or solution, as her description of them reveals.

My fourth, final category of the generative characteristic of the aesthetic relates to drawing on and working with intuition. In their interviews with me, Buhler and Borwein both described the way in which they could get, often quite suddenly, an 'out-of-the-blue' insight. They recognised it as an insight because of the strong feeling that accompanied it, almost alerting them to pay attention. [7] It may have some compelling order, simplicity or structure; it may resonate with something else they know; it may provide them with a new perspective. These are qualitative judgements they make. Buhler explained, "I have the idea but not the words". This convinces him – whether rightly or not – that it will lead to a solution. As Borwein said, a mathematician gets the "remarkable sense that the rest is do-able: this *will* work".

What are the types of things that make mathematicians feel that 'this will work out'? Very generally, they are things that have some aesthetic import. Hofstadter (1992, p. 5) senses the rightness of a particular relationship when he notices that it produces parallel lines – had the lines been oblique, he would have skipped right over them. He also feels that a simple analogy in symbolic form, though meaningless to him geometrically, *must* be right – such a thing *cannot* just be an accident. This 'looking right' is an elusive notion, one that stumps mathematicians who try to describe or explain it. Is there a perceptible harmony in terms of proportion or symmetry? Is there a resonance with a previously successful strategy? Are there simply some inexpressible or tacit conceptions that have finally found a formulation?

The third question here emphasises the aesthetic sensitivities that contribute to mathematicians' sense-making. In contemplating, experimenting with, playing with the elements of a situation, the mathematician is gaining a feel for patterns and potential patterns (see Martin Schiralli's discussion of pattern in Chapter 5 of this book). Hofstadter describes the sudden insight, the aesthetic moment, as being when inner images and external impressions converge; it is "the concrete realisation of the abstract analogy – a lovely idea, irresistible to me" (p. 7).

The mathematician may feel that she is bringing something beautiful but unfinished to its inevitable completion, to closure. In retrospect, she might appreciate the growth of her own grappling: she might be surprised (and thankful) that she pursued a certain path; she might realise how she wrongly dismissed something as irrelevant or meaningless along the way. This appreciation alerts her to the mysteries of her own mathematical thinking process, mysteries that in many ways parallel in their depth the mysteries she encounters in mathematics.

This generative characteristic of the aesthetic operates not only at a passive, sub-conscious level, as Poincaré would have it, but also actively, as the mathematician deliberately searches for order and structure. During this process, the mathematician becomes more intimate with objects and relations through various transformations and reformulations. As will become clear in the next section of this chapter, the generative characteristic of the aesthetic distinguishes itself in many different ways from the motivational one. While the latter pertains to what mathematicians perceive, the generative characteristic seems to relate to what they *do:* for example, playing around with or getting a feel for; gaining intimacy; using certain tools; calling on intuition. The act of *doing* in itself seems to carry a positive affective component with it, in the course of stimulating and supporting aesthetic modes of reasoning.

The Motivational Characteristic of the Aesthetic

Jacques Hadamard (1945), John von Neumann (1947) and Roger Penrose (1974) have all argued that that the motivations for doing mathematics, as Penrose states, "turn out to be ultimately aesthetic ones" (p. 266). Tymoczko (1993) claims that there is a logical imperative for the motivational characteristic of the aesthetic. A mathematician has a great variety of fields to choose from, widely differing from one another in character, style, aims and influence; within each field there is a variety of problems and phenomena.

Thus, mathematicians must select in terms of the research they pursue, the classes they teach and the 'canon' they help to pass on.

While there are some mathematical problems that are more famous and even more fashionable than others, it would be difficult to argue that there is an objective perspective – a mathematical reality against which the value of mathematical products can be measured. Contrast this with physics, for example, another discipline that makes strong aesthetic claims (see Curtin, 1982, or Farmelo, 2002). There, questions and products can be measured up against physical reality: for instance, how well they seem to explain the shape of the universe or the behaviour of light.

Hadamard (1945) firmly claimed that one of the most important motivational aesthetic criteria is that of *potential,* the fruitfulness of a future result:

> Without knowing anything further, we *feel* that such a direction of investigation is worth following; we feel that the question *in itself* deserves interest [...] (p. 127; *italics in original*)

When the young Le Lionnais perceived that initial symmetry, he could predict – or 'feel' – that the investigation would yield many treasures. An attraction to the potential of a result or to the harmony of a mathematical structure seems to appeal more to the intellect than to the senses. Penrose, however, describes another criterion, that of *visual appeal,* in explaining his attraction to the strange symmetries in his irregular tilings. Visual appeal seems to be an increasingly available criterion; the computer-generated images that are now being widely produced have bewitched many – as David Mumford *et al.* (2002) acknowledge in their recent and colourful book *Indra's Pearls.*

In analysing scientific inquiry in general, including mathematics, philosopher of science Michael Polanyi (1958) argued that the scientist's sense of intellectual beauty serves a crucial *selective* function:

> intellectual passions have an affirmative content; in science they affirm the scientific interest and value of certain facts, as against any lack of interest and value in others. (p. 159)

Moreover, Polanyi had already asserted in his book that the motivational characteristic of the aesthetic plays the specific psychological role that Penrose mentions above:

> Intellectual passions do not merely affirm the existence of harmonies which foreshadow an indeterminate range of future discoveries, but can evoke intimations of specific discoveries and sustain their persistent pursuit through years of labour. (p. 143)

Although the authors above argue that aesthetic motivation is necessary for mathematical inquiry, they provide very few examples of the types of aesthetic response that might be motivational. In the following sub-section, I provide categories of responses that are frequently mentioned in interview by mathematicians.

Categories of aesthetic motivation

There are several ways in which the aesthetic motivates mathematical activity: appeal, at both a sensory and cognitive level; surprise and paradox; orientation to the social; identification; desire for 'the feeling'.

Mathematicians can be attracted by the visual appeal of certain mathematical entities, by perceived aesthetic attributes such as simplicity and order or by some sense of 'fit' that applies to a whole structure. As I mentioned, Penrose (1974) is aesthetically motivated by the visual complexity of non-periodic tilings, but since so much of mathematics seems inaccessible to the senses, visual appeal is necessarily limited. Lewis points to another source of appeal in mathematics: "the unexpected order that so often emerges for no apparent reason" from complex situations. Davis (1997) provides a specific example, describing being caught by the unexpected order emerging from an irregular triangle in Napoleon's theorem and spending years of his life trying to figure out why it occurs.

Apparent simplicity is another frequently-occurring appeal and is exemplified by these words from Andrew Gleason (quoted in Albers *et al.*, 1990):

> I am gripped by explicit, easily stated things [...] I'm very fond of problems in which somehow an at least very simple sounding hypothesis is sufficient to really pinch something together and make something out of it. (p. 93)

Katherine Heinrich, one of my interviewees, concurred: "I like simply stated and clearly understood questions that with just a little background the 'man on the street' could understand", though she acknowledged that such problems are often "deceptively simple". Number theory [8] seems to attract many mathematicians, including Le Lionnais, as it swarms with problems and claims that are deceptively simple to state, such as Goldbach's conjecture or Fermat's Last Theorem.

In my interview with Hendrik Lenstra, he emphasised that his sense of attraction involved a sense of network, of connection and relationship: "It's the whole texture, the whole logical network that creates beauty". He compared his attraction to a mathematical situation with seeing "the insides of a watch, the beauty of the copper as well as of the rhythmic synchronisation of the motion of the gears". He explained that, "Something draws you in to look deeper, to see which movements are linked to which others or why a certain gear is there". Lenstra claimed that he will select questions based on his network view of mathematics; he will be attracted to a question if it "is not a dead end but is somehow connected to other things".

The mathematician's aesthetic response is necessarily personal, emerging from a certain set of preferences and interests. Osserman's source of attraction is partly the mystery of his own mind. He finds it remarkable that something he knows is true does not have a more transparent proof, that there is not a more "enlightening" explanation: if it is true, then he should be able to "see" it. The desire to 'see' it, to have an immediate gestalt of understanding is primarily an aesthetic one. It is not enough to know that something is true; one wants to be able to apprehend it in a holistic way. Similarly, Heinrich claimed that some problems generate on-going interest, because "no one has seen the 'right' way to do them". Such problems awaken a sense of desire to solve them in a more aesthetically pleasing way. (Witness the continuing attempts to find additional proofs of the irrationality of $\sqrt{2}$ – most recently in Apostol, 2000 – as described in Chapter 1.)

Secondly, a sense of surprise and paradox can also be aesthetically motivating. For example, the paradox of the 'hat problem' recently intrigued and attracted many mathematicians across North America (Robinson, 2001). Surprise constantly arises in mathematics, as mathematicians find things they have no reason to expect: a pattern emerging in a sequence of numbers; a common point of intersection found in a group of lines; a large change resulting from a small variation; a finite real thing proved by means of appeal to an infinite, possibly unreal, object. Movshovits-Hadar (1988) reveals the motivational power of surprise in mathematics, by showing how this feeling of surprise stimulates curiosity which can, in small steps, lead towards intelligibility.

Surprise makes one struggle with one's expectations, with the limitations of knowledge and, thus, with intuitive understanding, both informal and formal. Bill Gosper (quoted in Albers *et al.*, 1990) expressed surprise at the way continued fractions allow you to 'see' what a real number is: "it's completely astounding [...] it looks like you are cheating God somehow" (p. 112). He claimed this sense of surprise had motivated him to do extensive work with continued fractions. Of course, in order to respond to surprise, one must have some kind of frame of reference that generates expectations, so that something that surprises one person may not surprise another (see also Stanley, 2002).

The work on foundations of mathematics provides a good example of non-surprising problems, almost by definition. In fact, Krull (1930/1987) suggested that those attracted to the study of foundations (investigating, for example, the extent to which the set of all infinite decimals can be considered a logically faultless concept) are the least aesthetically oriented mathematicians. He argued this is so since they are "concerned above all with the irrefutable certainty" (p. 50) of their results. He additionally claimed that:

> the more aesthetically oriented mathematicians will have less interest
> in the study of foundations, with its painstaking and often necessarily
> complicated and unattractive investigations. (p. 50)

Krull quite clearly situated himself in the latter camp, but was perhaps too quick to judge 'foundations' mathematicians as a whole as non-aesthetically oriented. The inclination toward finding basic, underlying order is certainly an aesthetic one – though different in kind from the inclination to surprise. This differentiation in motivation evokes Le Lionnais's distinction between 'classical' and 'romantic' impulses.

The response of surprise can sometimes be oriented toward the mathematician's own way of thinking, rather than toward the mathematics itself. For example, Lenstra recalled his amazement at the fact that he can much more readily understand and "control" infinity than the enormous numbers – known as Ramsey numbers – required in a combinatorial proof that he worked on. "Isn't that beautiful", he exclaimed, at the apparent paradox of being able to understand the "biggest but not the very big". Similarly, Buhler found pleasure in seeing a new feature of a phenomenon he thought he knew so well. It surprised him. "I couldn't believe that I hadn't seen it before. It was so basic." This was the hook that initiated his mathematical inquiry. He had to work out why this feature – so obvious as it seemed now – had previously eluded him: "How *ever* could I have missed it?"

Thirdly, there is also a social dimension to aesthetic motivation. William Thurston (1994) agrees with Penrose, Hadamard and von Neumann on the necessary aesthetic dimension to a mathematician's choice of field and problems, but he adds another, one that is rarely discussed:

> social setting is also important. We are inspired by other people, we seek appreciation by other people and we like to help other people solve their mathematical problems. (p. 171)

In my interview with him, Lenstra concurred:

> There is not much fun in deciding for yourself that a particular area of mathematics is beautiful and spend your life on it if you are the only person who finds it beautiful. If someone else is interested in something that you are doing, that's an enormous boost. It's a real stimulus.

For Buhler, a judgement of significance can be affected by social influences – colleagues can convey the promise of pleasure through "infectious excitement".

These observations provide some indication of how mathematicians' aesthetic choices might (partially, at least) be learned from their community as they interact with other mathematicians and also seek their approval. Of course, not all social interactions among mathematicians have an aesthetic dimension. The case of John Nash exemplifies a non-aesthetic social motivation. His biographer Sylvia Nasar (1998) describes how he would only work on a problem once he had ascertained that great mathematicians thought it highly important – pestering them for affirmation. The promise of recognition, rather than the intrinsic appeal of the problem or situation, seemed to be the motivating factor.

A fourth source of motivation might come from a sense of *identification,* when a mathematician perceives a rapport – a connection between the situation and her own interests and aptitudes. As Osserman explained in his interview with me, he is drawn to a problem when he realises that it is like, or connects to, something that he already knows. And this is so, regardless of whether it is a concept or method of mathematics and whether or not it is an approach or style of his own. Buhler claimed he is attracted to a problem if "it relates to something that I've done".

Lenstra was even more candid, explaining that he would only attack a problem "if it is the type of problem I am good at solving". Gleason (in Albers *et al.,* 1990) explained that he is specifically attracted to problems that "go from what might be called a qualitative way of looking at things to the quantitative way of looking at things" (p. 91), for instance problems that combine classical and analytic geometry. A mathematician's judgement of rapport can provides a sense of confidence and the pleasure of knowing that she is particularly well-suited to solving a problem, that there is something about the way she thinks that fits with the mathematics, that the problem is – in a way – amenable to becoming hers.

My fifth and final motivating factor is different in kind from the others. In all the cases described above, the aesthetic basis for motivation depended upon perceiving the qualities of the problem or situation itself. However, the motivating factor of longing for 'the feeling' depends more on the mathematician's prior, positive experiences of mathematical inquiry. Thus, a mathematician might embark on a particular exploration or problem-solving process because she remembers the feelings of tension, of puzzlement, of frustration and of final satisfaction that make up her successful mathematical experience. She explicitly seeks out this experience when selecting certain problems to work on.

There are many periods of drudgery in mathematical work: as Lewis remarked, "You often have to do relatively grungy mathematics, which isn't terribly appealing, but it is part of what you are doing". However, a mathematician keeps returning for those special moments when, as Heinrich explained, "I see something of elegance and almost magic – when you think that's how a proof is meant to be, or you are just amazed that a particular thing could be true". Lenstra claimed that such feelings are what keep him doing mathematics: "It is a very good feeling, I wouldn't do mathematics if it weren't for that". Mathematicians can also have this feeling as 'spectators' rather than creators. Heinrich recalled the moving experience she had when a colleague presented a new proof of an old result at a conference: "It was so beautiful and elegant everyone spontaneously applauded when they saw how it was working – somehow everyone knew they had just seen something significant".

Polanyi (1958) insisted that the various ways in which mathematicians become attracted to mathematical situations and problems do not solely serve an affective motivational purpose. Rather, the attraction also has a *heuristic* function, by influencing the ability to discern features in a situation and thereby directing the thought patterns of the inquirer. He suggested that the motivational characteristic of the aesthetic does not operate merely as an 'eye-catching' device, nor does it provide merely the psychological support needed to struggle through a problem. It is also central to the very process that enables the mathematician to produce qualitatively derived hypotheses deliberately: it initiates an action-guiding hypothesis.

Some Concluding Remarks

Most surface definitions of mathematics describe the materials mathematicians work with: 'mathematics is the study of shapes and numbers', 'mathematics is a description of nature' or 'mathematics is a theory of formal patterns'. An analogous definition for the visual arts – so obviously inadequate – might read: 'visual art is the manipulation of form, colour and texture'.

Some less objective definitions of mathematics remind that mathematics is a tool we use to interpret the world: 'mathematics is a language'. But few purported definitions capture the animating purposes of mathematicians: *why* do mathematicians do mathematics? What impulses, what inclinations are responsible for producing the body of knowledge that is mathematics?

This chapter has at least begun to answer these questions, identifying the aesthetic basis for many of the choices that mathematicians make when posing, solving and sharing problems. I close with the following definition of mathematics, offered by James Shaw:

> Mathematics is, on the artistic side, a creation of new rhythms, orders, designs and harmonies, and on the knowledge side, is a systematic study of the various rhythms, orders, designs and harmonies. Mathematics is, on the one side, the qualitative study of the structure of beauty, and on the other side is the creator of new artistic forms of beauty. (in Schaaf, 1948, p. 50)

For me, this attempt at a definition subtly suggests the ways in which mathematics satisfies the basic human impulse to find and describe pattern.

Notes

[1] This assertion stands in contrast to what Hardy (1940) implied in his discussion of mathematical beauty, as well as in contrast to the traditional conception of aesthetics found in philosophy and art criticism (see, for example, Bell, 1914/1992, pp. 113-116). However, the survey conducted by Wells (1990) provides substantial evidence that not all mathematicians share the same aesthetic values; that their experiences and preferences, as well as their states of mind, may greatly affect their aesthetic judgements.

[2] In the past, some mathematicians have referred to Euler's equation ($e^{i\pi} + 1 = 0$) as one of the most beautiful in mathematics, but others nowadays think it is too obvious to be called beautiful (Wells, 1990). Schattschneider (see Chapter 2) still finds this formula beautiful. When assessing a mathematical idea's aesthetic value, she might agree with the mathematician in Wells's survey who wrote, "I tried to remember the feelings I had when I first heard of it" (p. 39).

[3] Wolfgang Krull (1930/1987) suggests a very similar line of division: the *concrete* (instead of the romantic) versus the *abstract* (instead of the classical). He sees mathematicians with concrete inclinations as being attracted to "diversity, variegation and the like", comparing them with those who prefer heavily ornamented buildings. On the other hand, those with an abstract orientation prefer "simplicity, clarity, and

great 'line'" (p. 52). And Freeman Dyson (1982, pp. 49-55) has suggested a related line of division in the sciences, distinguishing scientific 'diversifiers' (e.g. Rutherford) from 'unifiers' (e.g. Einstein). Unifiers use "the enormous power of mathematical symmetry as a tool of discovery" (p. 50) and they are "happy if they can leave the universe looking a little simpler than they found it" (p. 51). Diversifiers are symmetry-breakers who are "happy if they leave the universe a little more complicated than they found it" (p. 51). Moreover, Dyson claimed these two types are complementary in the quantum-theoretical sense. He went on, "It is easy to understand why we have two kinds of scientists, the unifiers looking inward and backward, the diversifiers looking outward and forward into the future" (p. 51). Dyson concluded: "every science needs for its healthy growth a creative balance between unifiers and diversifiers" (p. 54).

[4] It may seem odd to classify magic squares and Pascal's triangle as mathematical 'facts'. In choosing this term, Le Lionnais may have been attempting to distinguish static mathematical ideas, which are the result of mathematical inquiry, from the processes (or 'methods') that generate those ideas.

[5] I interviewed Jonathan Borwein, who until 2004 led the Centre for Experimental and Constructive Mathematics at Simon Fraser University, Canada, as part of a larger study into the roles of the aesthetic in the activities of contemporary mathematicians (Sinclair, 2002). I also interviewed the following mathematicians: Joe Buhler (Reed College), Katherine Heinrich (University of Regina), Hendrik Lenstra (University of California, Berkeley), Adrian Lewis (University of Waterloo) and Robert Osserman (MSRI). In order to increase this chapter's readability, as well as to distinguish quotations by these mathematicians arising from my interviews from others found in the literature, excerpts from my interviews will be used throughout this chapter without further reference being given. The first occasion I quote from these interviewees, I use their full name; subsequent to that, I just refer to them as, for example, Borwein or Heinrich.

[6] This comparison between poetry and mathematics has in fact been noted by several scholars – e.g. Gösta Mittag-Leffler compared the works of the mathematician Niels Henrik Abel with "truly lyrical poems possessing a supreme beauty, in which the perfection of form reveals the depth of thought" (quoted in Le Lionnais, 1948/1986, p. 456; *my translation*). Also, most recently, see Mazur (2003).

[7] See Richard Skemp's (1979) *Intelligence, Learning and Action* for a more in-depth discussion of the role of emotions in problem solving.

[8] André Weil (1984) suggests that number theory may, in fact, surpass all other fields in its quantity of deceptively simple problems. The simplicity may arise, at least in part, from the fact that its primary objects – whole numbers – are among the most basic and familiar ones in mathematics.

CHAPTER 5
The Meaning of Pattern

Martin Schiralli

In the late 1970s, when the eminent anthropologist and biologist Gregory Bateson sought to codify his influential views on the ecology of mind, he chose the idea of pattern as his central heuristic device. The choice was not surprising, for Bateson, in a remarkably productive career as both scientist and educator, had by that time been using this concept to explore, identify and represent the essential features of biology and anthropology for more than a quarter of a century. In his summative *Mind and Nature,* published in 1979, Bateson related one early experience in his career that illustrates particularly well the power that the notion of pattern can have in helping to organise one's thinking in fundamental ways.

In the 1950s, Bateson was teaching a course aimed at introducing the essential purposes of biological science to a group of art students at the California School of Fine Arts in San Francisco. Astutely realising that the group of about a dozen students would be particularly responsive to the visual and the tactile, Bateson came to the initial class with a paper bag containing the remains of a recently-cooked crab. Putting the crab on the table in front of the students, he asked them to produce arguments in support of the contention that the thing before them was indeed the remains of a living thing. As Bateson had correctly anticipated, the students' attention was drawn to the perceptual qualities of the crab. An animated discussion ensued during which the idea of symmetry was offered by some students and rejected by others as suitable proof of the claim.

The point at issue was that while there was admittedly an ordered correspondence between the claws, one claw was significantly larger than the other, which thereby made talk of symmetry problematic. Finally, one student said, "Yes, one claw is bigger than the other, but both claws are made of the same parts" (p. 9). In recalling the moment, Bateson was delighted with this observation:

> Ah! What a beautiful and noble statement that is, how the speaker politely flung into the trash can the idea that *size* could be of primary or profound importance and went after the *pattern which connects.* He discarded an asymmetry in size in favor of a deeper symmetry in formal relations. (p. 9; *italics in original*)

In Bateson's demonstration, patterns within a specific, individual crab (first-order connections) would soon give rise to second-order patterns, those phylogenetic homologies that connect crabs to, say, lobsters. Finally, connections that link crabs and lobsters also suggest third-order patterns between other homologies like that between horses and men. In this way, Bateson and his students constructed "a ladder of how to think about [...] the pattern which connects" (p. 11).

Reflecting on his success with the art students, Bateson identified one significant element in their own backgrounds and aptitudes:

> I faced them with what was (though I knew it not) an aesthetic question: *How are you related to this creature? What pattern connects you to it?* [...] I [...] forced the diagnosis of life back into identification with living self: "*You* carry the bench marks, the criteria, with which you could look at the crab to find that it, too, carries the same marks." My question was much more sophisticated than I knew. (p. 9; *italics in original*)

The identification of a pattern is, therefore, a fundamentally aesthetic apprehension that in systematic inquiry soon moves beyond the immediately perceptible towards the more formal conceptual connections with which scientific and mathematical theory is ultimately concerned. Thus, in Bateson's mature thought, the 'pattern which connects' became the meta-pattern, the dynamic sub-stratum of purposeful structures and functions – those interconnecting, evolving lines of functional symmetries and correspondences that link all living things, including human minds.

For my present purpose, that of gaining some points of purchase on the concept of 'pattern' itself, Bateson's account provides a vivid illustration of the way in which this concept may function comfortably at or among the levels of aesthetic perception, empirical investigation and formal relation. One further element in this account is of particular significance. Bateson cautioned his readers against having too rigid a conception of 'pattern':

> We have been trained to think of patterns [...] as fixed affairs. It is easier and lazier that way but, of course, all nonsense. In truth, the right way to begin to think about the pattern which connects is to think of it as *primarily* [...] a dance of interacting parts and only secondarily pegged down by various sorts of physical limits and by those limits which organisms characteristically impose. (p. 13; *italics in original*)

One might observe that it is not only by training that we have come to think of pattern in this way, for 'pattern' in one of its senses, namely the fixed and easily replicable pattern of the cookie-cutter or, perhaps, the arithmetic progression, is quite determinate. The conceptual difficulty of moving from this sense to the related, more indeterminate "dance of interacting parts" that Bateson rightly saw as a more powerful sense of the term will be recognised immediately upon filling in the ellipsis immediately following the word *'primarily'* in the above quotation. For Bateson wrote "*primarily* (whatever that means) a dance of interacting parts". This chapter will later attempt to provide a way of making sense of Bateson's use of the word 'primarily' in this connection.

At about the same time that Gregory Bateson's *Mind and Nature* appeared, the art historian Ernst Gombrich was preparing for publication his own study of pattern in visual art. Having earlier presented the results of his inquiries into the psychology of perception in representational art, in the definitive *Art and Illusion* (1960), Gombrich was now poised to explore the still more subtle problems of the psychology of abstract design and the art of decoration. Substantially reworking and expanding the drafts of the 1970 Wrightsman lectures, which he had originally given in New York, Gombrich's *The Sense of Order* was, like Bateson's *Mind and Nature,* published in 1979.

In this comprehensive study, Gombrich located the animating principle for abstract and decorative art in a human need to find and to consolidate patterns in experience. Influenced by broad intellectual currents as diverse as Immanuel Kant, Karl Popper and Konrad Lorenz, Gombrich provided an account of pattern grounded in epistemology and ethology, as well as in psychology. For Gombrich, the sense of order was closely connected to Popper's notion of the need for regularity. In *Objective Knowledge,* Popper (1972) had written:

> It was first in animals and children, but later also in adults, that I observed the immensely powerful *need for regularity* – the need which makes them seek for regularities [...] (p. 23; *italics in original*)

Gombrich (1979) maintained that this need for regularity is the product of part of the instinctual scaffolding that all organisms possess, a "built-in hypothesis" (p. 3) respecting the possibility of locating and exploiting environmental regularities in the on-going business of living. The mode of hypothesis confirmation, or better, refutation – for Gombrich was a genuine Popperian – is perception itself, now functioning metaphorically as a searchlight. In a particularly telling passage, Gombrich wrote:

> the 'searchlight' metaphor comes in useful, for it reminds us of the activity that is inseparable from the most primitive model of perception. The organism must probe the environment and must, as it were, plot the message it receives against that elementary expectation of regularity which underlies what I call the sense of order. (p. 3)

For Gombrich, this sense of order is at work when we *discern* patterns, as well as when we make them. The human propensity to make decorative patterns, the special relationship of patterning to mathematics in general (and to geometry in particular) and the intentions and tools of the pattern-makers are surveyed in Gombrich's generative study. The effects of such deliberately contrived instances of "ordered profusion" (p. 16) on the human mind, their relationship to the more representational figures of traditional iconography (see also Chapter 9 in our book) and even an extrapolation into the realm of auditory phenomena and music were likewise explored in the remainder of this important work.

Like Bateson, Gombrich acknowledged a proclivity in thinking about patterning towards the fixed, readily progressive patterns we all acknowledge as central cases of the concept. But, also like Bateson, Gombrich saw the importance of relating the concept to the more open and less easily specified ordered arrangements that may, in Bateson's image, 'dance' for us in more complex contexts:

> Here the sense of order is given free rein in generating patterns of any degree of clarity or complication. We cannot prescribe to the designer whether he should aim at restlessness or repose. The West generally preferred symmetry, the Far East more subtle forms of balance. (p. 146)

Still other patterns, Gombrich continued, "explore the instability derived from the wealth of different interpretations the design offers to the searching eye" (p. 160). It may be noted that this brings us very close, of course, to the world of the abstract fine artist whose works may often display such an ordered instability. This is a kind of patterning whose inner logic resists description in terms of readily identifiable symmetries and regularities, but which nonetheless achieves an *aesthetic* stasis of unity and purposeful integrity.

Although approaching the idea of pattern from the ostensibly dissimilar subject matters of visual art and biology, both Gombrich and Bateson may be seen to have opened the concept of 'pattern' for use as an intellectual tool for discerning and representing less than fully determinate regularities. The abstract design of rectilinear planes in a painting by Piet Mondrian or Theo van Doesburg and the apparently haphazard splashes of paint on a Jackson Pollock canvas may, likewise, both be perceptually interrogated in terms of pattern (see Walter, 2001; Pimm, 2001; Taylor, Micolich and Jones, 1999) – just as Bateson's homological correspondences may be conceptualised within the same connecting pattern as the human mind itself.

Beauty Bared

> A mathematician, like a painter or a poet, is a maker of patterns. If his patterns are more permanent than theirs, it is because they are made with *ideas*. A painter makes patterns with shapes and colours, a poet with words. [...] A mathematician, on the other hand, has no material to work with but ideas, and so his patterns are likely to last longer since ideas wear less with time than words. (Hardy, 1940, pp. 84-85; *italics in original*)

In asserting that poets, painters and mathematicians are all makers of patterns, Cambridge analyst and number theorist Godfrey Harold Hardy identified a common element between mathematics and the arts. If the beauty arising from the patterns of verbal and visual materials in art is more palpable, Hardy further proposed, the intellectual beauty of mathematics is of a far more durable kind. In choosing to focus on the relative longevity of the two types of beauty, rather than their respective and comparative values, Hardy here resisted the temptation at least as old as Pythagoras to find in the beauty of mathematics the very highest order of aesthetic interest possible.

Bertrand Russell, however, had no difficulty in considering the beauty of mathematics to be "supreme". Considered correctly, Russell (1917) maintained, mathematics may be seen to possess "a beauty cold and austere, like that of sculpture". But this "sublimely pure", "stern perfection" has no "appeal to any part of our weaker nature". It is "without the gorgeous trappings of painting or music" (one is almost tempted to read "uncontaminated by" for "without") and routinely produces a beauty that "only the greatest art can show" (p. 57).

Perhaps the finest expression of this strong view of mathematical beauty occurs, somewhat ironically, in that more fragile medium, poetry. Edna St. Vincent Millay's (1956, p. 605) sonnet, *Euclid alone has looked on beauty bare* – from which the heading of this section is derived – is well-known and much admired by mathematicians. This admiration is not too difficult to understand, for it is indeed a fine poem; nor does the poet's presentation of the mathematician in passionate, grandly heroic terms create too many textual problems for this group of readers. Indeed, the poem is almost mathematical itself in the elegance, lucidity and economy of means by which the special beauty of mathematics is demonstrated. Let those who merely "prate of Beauty hold their peace", Millay urged, for theirs is the mundane perceptual sphere of "dusty bondage" where "geese gabble and hiss" in muddled confusion about the true nature of aesthetic value.

Happily, those distorted views of beauty, dependent as they are on the frailties of light reflecting on mutable things, have been transcended by Euclid in whose *Elements* light itself is "anatomized". This human glimpse of "light anatomized" or beauty liberated from the constraints of perceptual distortion is indeed a glimpse of Aphrodite, the goddess of Beauty herself, whose "massive sandal set on stone" was felt only at the moment of Euclid's inspiration. Empowered by this powerful vision, succeeding generations of mathematicians may as "heroes seek release [...] into the luminous air".

As the substance of poetic value here is verbal expression, one may find an additional irony. This poem provides a vivid mythopoeic description of the unique beauty possessed by mathematics, one that would be impossible to provide, Bertrand Russell notwithstanding, in the admittedly more pristine terms of mathematical expression itself.

'Pattern' in Art and Mathematics

Among mathematicians, therefore, the notion of pattern is frequently used in describing the essential core of their activity. To Hardy's and Russell's may be added the voice of Warwick Sawyer (1943, 1955, 1970) who wrote extensively on this theme. The interpenetration of aesthetic and formal considerations in these descriptions, an approach I have already shown to characterise Gombrich's view of art and Bateson's view of nature, has also played an important part in the long history of mathematicians' thinking about mathematics.

Also among mathematicians there is a growing interest in the less-fixed possibilities of pattern that Gombrich and Bateson alerted us to. Although one might be tempted to see less scope for the notion of indeterminate pattern in a field in which inexorable proofs are pursued so relentlessly, and on occasion with so much drama, contemporary mathematicians would be quick to agree that indeterminate or 'fuzzy' patterns are now a vital part of mathematical inquiry. The mathematical fact that the number π is not algebraic but transcendental, for example, does not preclude the search for 'themes' within and among the myriad digits stretching out seemingly indeterminately beyond the decimal point (see Chapter 1 for more detail).

The contemporary mathematician Keith Devlin (1994) has traced the roots of the mathematician's proclivity to talk of pattern to the ancient Greeks, for whom mathematics had not only intense intellectual interest but aesthetic value and spiritual significance as well. By focusing on the activities of mathematicians, including historical examples of the doing of mathematics, his book *Mathematics: the Science of Patterns* is more than an introduction to the subject matter of mathematics as formally represented. It is an attempt to capture the very essence of mathematical inquiry in terms of the concept of pattern itself.

Devlin characterises each branch of mathematics as the exploration of pattern: the patterns of number and counting as the subject matter of number theory, while geometry studies patterns of shapes. Devlin also identifies those patterns of reasoning that underlie mathematical logic, while those of motion form the subject matter of calculus. Patterns of position and closeness comprise the study of topology and probability theory attends to patterns of chance.

If by 'science' of patterns, however, Devlin means the systematic exploration and representation of pattern possibilities, it can easily be shown that art likewise explores pattern possibilities. These explorations are often quite systematic (as, for instance, in the case of Monet's studies of Rouen Cathedral under a range of daylight conditions, the mosaics of the Alhambra mosque in Granada or Bach's *Art of the Fugue*) and these possibilities are likewise represented in publicly accessible forms.

One might counter that the artist is more concerned with the 'creative' exploration of pattern possibilities than the mathematician, but few mathematicians would fail to acknowledge the central role that creativity plays in genuine mathematical activity. Providing a more detailed characterisation of the various activities routinely undertaken by mathematicians when doing or studying mathematics may help to clarify the issue, particularly if these activities are compared with those routinely practised by people making or studying art.

Devlin is certainly right in claiming that mathematicians are intrigued by patterns. Mathematicians often:

- *wonder* at the patterns discerned in experience;
- *analyse* patterns – noticing, noting, associating patterns and elements of patterns;
- *represent* patterns, i.e. describe them in formal terms;
- *manipulate* patterns;
- *create* novel or original symbolic patterns;
- *imagine* the possibilities of patterns;
- *connect* pattern possibilities, i.e. analyse, classify and theorise patterns, thereby creating larger, more comprehensive patterns.

Mathematicians also:

- *demonstrate,* i.e. prove (or describe) the necessity (or nature) of patterned relationships using other patterns, viz. the patterns of logical operations.

In so doing, mathematicians:

- *compute,* i.e. perform operations on patterned relationships using other patterns, *viz.* arithmetical, algebraic, and so forth.

Finally, mathematicians:

- *appreciate* the historical and contemporary achievements of other mathematicians;
- *evaluate* the achievements of other mathematicians.

While there is no gainsaying that mathematicians study patterns, many of the activities specified above may also be noted in the arts. Although their interest also embraces other aspects of perceptually interesting phenomena, artists or scholars in the arts are likewise intrigued by patterns. They:

- *wonder* at the patterns, sights, textures, sounds and apparent emotional vitality of perceptual phenomena;
- *analyse* patterns: noticing, noting, associating patterns and elements of patterns;
- *represent* patterns, i.e. embody them in sensory terms;
- *manipulate* patterns using different media and materials;
- *create* novel or original expressive patterns;
- *imagine* the possibilities of patterns;
- *connect* pattern possibilities, i.e. conceptualise and theorise patterns creating larger, more comprehensive patterns.

Artists or arts scholars also:

- *attempt* to understand the nature and possibilities of patterned relationships using other patterns, viz. those of logical operations.

In so doing, artists or arts scholars also:

- *develop* stylistic 'vocabularies' of patterns, using ideas, media, techniques and materials, as well as other aspects of aesthetic interest such as sensory and expressive qualities;
- *appreciate* the historical and contemporary achievements of other artists and arts scholars;
- *evaluate* the achievements of other artists and arts scholars.

At one level, therefore, it would seem that certain root activities that are common to both domains may be detected in the spheres of both practicing mathematicians and practicing artists or arts scholars. Although the ends or purposes of these common activities are obviously distinct when undertaken in either a mathematical or an artistic context, they nonetheless share some very significant core components.

In doing calculus, for instance, it may be said that the mathematician is exploring patterns in motion and time. But composers, choreographers, film-makers, poets and other artists likewise explore patterns in motion and time. Of course, in doing calculus, the mathematician does many other things the artist does not do and the artist likewise does things in the arts that the mathematician does not do: but at one level they do nevertheless both explore patterns in motion and time. Similarly, in making a photograph or an oil painting, the artist explores patterns of shapes and surfaces, of edges and proximities – activities not so unlike those of the geometer and the topologist.

Although the ends or purposes of mathematical and artistic activities are distinct – with the mathematical centrally concerned with manipulable symbolic representations and the artistic centrally concerned with representations in sensory terms – underlying those differences some significant commonalities may be detected. In so far as actively engaging the subject matter of patterns is concerned, the arts would appear to have as strong a proprietorial claim on the concept as mathematics.

In defining mathematics in terms of the subject matter of patterns, therefore, it would seem that Devlin's specification raises some important questions. At the actual level in which the commonalities discussed above occur, it would appear that the mathematical and the aesthetic are both embedded in a very *special* relationship outside Devlin's definition (or better, perhaps, *underneath* it), one that warrants further analysis. Happily, in reminding us very early in the book of the proclivity of mathematicians to define themselves and their inquiries by means of the language of patterns, pattern-making and beauty, through the words of Hardy and Russell already quoted above, Devlin has oriented his discussion within a tradition that reaches back to the very origins of systematic mathematical demonstration among the ancient Greeks.

In so doing, and especially in view of his reliance on 'pattern' as the essential component of mathematical study, Devlin invites a reconsideration of those ancient issues surrounding the nature of the mathematical that may help in answering the questions provoked by the ease with which art and the aesthetic insinuate themselves into his definitional stance on mathematics. I will, therefore, pursue further the principal sources of the Greek, especially the Pythagorean, preoccupation with the interconnectedness of the mathematical and the aesthetic – through the mediating force of patterns.

Pythagorean Pattern

Little is known about the historical Pythagoras, apart from the fact that he was born towards the middle of the sixth century BC on the island of Samos. An Ionian Greek, he emigrated to southern Italy where he founded a society with religious, philosophical and possibly political interests. That society, the *Order of the Pythagoreans,* was secret and it is probably for that reason that little was actually written down by the Pythagoreans themselves. By the middle of the fifth century BC, the members of the society had dispersed throughout the Greek-speaking world and their founder had become the object of considerable legend and lore (see, for example, von Fritz, 1975).

With one significant exception – the fragments of Philolaus of Croton – whatever accounts we do have of Pythagoras and the Pythagoreans, especially with regard to mathematics, are often tendentious. These descriptions were written by later philosophers unsympathetic to (their versions of) Pythagorean doctrines (Aristotle and the Aristotelians) or by others (Plato and the Platonists) whose elaboration of these doctrines moved them into philosophical systems with significantly different underlying presumptions from those the Pythagoreans may actually have held themselves.

It has generally been considered reasonable, however, to ascribe to Pythagoras and the earliest Pythagoreans at least the following:

- the mathematical theorem bearing Pythagoras's name, equating the sum of the squares on the two shorter sides of a right triangle and the square on its hypotenuse;
- the discovery that $\sqrt{2}$ (the ratio of the side and diagonal of a square) is 'irrational' (*alogos,* i.e. indeterminate);
- the introduction of the doctrine of metempsychosis (the transmigration of souls) into Greek thinking, a view respecting immortality that originated prehistorically in Indian thought and culture;
- a mystical apprehension of the pervasiveness of number and harmony in the substance and structure of the universe (including what we would now separately identify as mathematics, music theory and astronomy).

The first two items have more directly mathematical import. With regard to an apparently passing detail in the first item in this list, the difference between the preposition 'on' (as used here) and the now more common preposition 'of' reflects a world of mathematical difference.

Squares 'on' sides of triangles signal a perspective whereby the figures that are provably 'equal' are geometric squares (and such proofs often employ finite dissection methods, discussed briefly in Chapter 2). The preposition 'of' reflects a twenty-four-hundred-year later sensibility, whereby lengths are taken to be real numbers and the theorem involves the arithmetic operation of numerical squaring being applied to them (see Fowler, 1985a, 1985b). Richard Dedekind (1872/1963) himself wrote about his sense of the significant challenge of establishing secure proofs for results involving arithmetic operations with irrational numbers:

> Just as addition is defined, so can the other operations of the so-called elementary arithmetic be defined, viz., the formation of differences, products, quotients, powers, roots, logarithms, and in this way we arrive at real proofs of theorems (as, e.g., $\sqrt{2} \cdot \sqrt{3} = \sqrt{6}$), which to the best of my knowledge have never been established before. (p. 22)

Likewise, in relation to the second item in the earlier list, many refer (as I deliberately did here) to the Pythagorean result as 'proving that $\sqrt{2}$ is irrational'. However, to do so implies a dramatically different conception of number, immeasurably far from that which the best sources available (admittedly few and far between) could remotely justify.

With regard to the latter two items listed earlier, whether viewing the spiritual sphere as an ordered arrangement of appropriately recycled souls passing successively, and fittingly, through various forms of life or viewing reality as ultimately to be revealed in mathematically harmonious concepts, the notion of pattern permeated Pythagorean thinking.

Number as pattern

A key to understanding the significance of pattern among the Pythagoreans is to be found in the actual numerals used by the early Greeks, for in considering the notational system used by these ancient mathematicians, the distinctiveness of the early Greek conception of number may be inferred. There were in fact two very different systems of numerical notation at work among the ancient Greeks. Karl Menninger's (1958/1969) *Number Words and Number Symbols* informs us that the earlier of the two, "arranged the numbers in order and grouped them like the Roman numerals" (p. 268).

> [These so-called] row numerals [...] are patterned on a decimal 10-grouping interrupted by a quinary 5-grouping. The units are represented by vertical strokes. (p. 268)

In the fifth century BC, there appeared a new, more "erudite system of alphabetical numerals" (p. 268). But this system "was not adopted as the official system of numerals in Athens until the 1st century B.C." (p. 268). The Greek alphabetical numerals made use of the twenty-four letters of the Greek alphabet augmented by three more. The first nine stood for 1 to 9, the next nine for 10 to 90 and the final nine for 100 to 900. A comma was used to mark the 'thousands' place. (In passing, note how such double duty for letters as components of words and as 'digits' of numbers certainly makes for a greater plausibility for numerology.) One disadvantage of this non-place value system compared with our current place-value system is that there is no symbolic link among, say, the symbols for 3, 30 and 300 to exploit algorithmically. However, perhaps a more pertinent question is to ask in what ways were the new numerals to be preferred over the older ones?

Their advantage over the old row numerals can hardly be missed;
they represent an enormous simplification, since they use only one
sign for each unit (rank) [...] another advantage of these alphabetical numerals, in fact their most important advantage: With these
numerals it was possible at long last to make computations in writing, without having to use an abacus. (Menninger, pp. 271-272)

While the advantage of written computation with numerals alone, as
opposed to manipulative work with physical representations of numbers,
has frequently been overestimated (see Rotman, 1987, or Tahta, 1991), it certainly altered the relationship of the user to the trace of 'ordered plurality'
(see below) that resided in the older numerals. While the alphabetical
numerals permitted more robust written calculations, they also obliterated
the underlying iconic pattern that in the older system linked numerals with
palpably isomorphic configurations of discretely ordered units and groups
of units. What the older numerals also provided by means of their configured patterning of quantities was something quite important to the early
Greek and Pythagorean conception of number.

Arithmos

In his landmark work, *Lore and Science in Ancient Pythagoreanism,* Walter
Burkert (1972) cautioned that *arithmos* (pl. *arithmoi*), the early Greek term
widely used for number, should not be confused with more modern conceptions of number:

> Αριθμος [*Arithmos*] is always a whole number, and tied up with the
> actual procedure of counting. Thus it is closely connected with
> things, and in fact is itself a thing, or at least an ordering of things.
> Αριθμος means a numerically arranged system, or its parts. (p. 265)

This distinction has been well acknowledged in the traditions of classical
scholarship and Burkert noted a somewhat earlier and quite vivid account by
Oskar Becker (1957), who rendered *arithmos* into the German term *geord-
nete Mannigfaltigkeit* ('ordered plurality') (pp. 21-22). He compared this conception with related contemporary notions like 'couple', 'dozen' and 'score'.

Subsequently, the historian of mathematics David Fowler (1999) claimed:

> a much more faithful impression [than cardinal numbers] of the very
> concrete sense of the Greek *arithmoi* is given by the sequence:
> duet, trio, quartet, quintet, ... [...] These numbers are ordered by
> size ('a quartet is bigger than a trio'), and can be added by concate-
> nation ('a trio plus a quartet makes a septet') and subtracted 'the less
> from the greater'. The *arithmoi* also appear in other forms, such as
> the adverbial sequence: once, twice, three-times, four-times, ... [...]
> I shall refer to them as 'repetition numbers'. (pp. 13-14)

Viewing number, as the Pythagoreans must surely have done, as ordered
plurality shows how central the notion of ordered arrangement or pattern
was to their conception of number. It also helps explain a number of

Pythgorean positions, such as their view that the universe is revealed in numbers (discernible patterns or ordered arrangements), the intense interest in the collateral geometric and arithmetic branches of mathematics (as evidenced, for instance, by the fascination with the properties of 'polygonal' numbers, as well as the phenomenon of incommensurability) and, indeed, their apprehension of the mystical properties of numbers-in-things.

This serious mathematical and spiritual interest in number as pattern is indicated by the fact that the Pythagoreans "devoted great efforts to the study of figurate numbers: triangular, square, pentagonal, and so on" (Gazalé, 1999, p. 11). Indeed, such figurate numbers were "represented geometrically by triangles of stones laid on the ground" (Grattan-Guinness, 1997, p. 46). The *tetraktys,* the pattern representing the fourth of the triangular numbers, 10, was "deemed so mystical by Pythagoras's followers that they adopted it as the emblem of their secret brotherhood" (Gazalé, 1999, p. 12).

However, there was plausibly also a far more subtle numerical patterning through *arithmoi* at work, arising in the context of incommensurability. Historian Ivor Grattan-Guinness (1997) has provided an account of the purported Pythagorean discovery with regard to incommensurability that will further help in understanding these fascinations:

> Another famous finding attributed to the Pythagoreans is usually formulated thus: *The number $\sqrt{2}$ is irrational;* but this formulation is anachronistic in various ways. Firstly, "(ir)rational" have become normal adjectives in European languages, due to Latin translations; but they give a wrong impression, and the Greek words "(a)logos" are better rendered as "word(less)", and "ar(rhetos)" as "(in)expressible". Secondly, the theorem concerns numbers, whereas when the Greeks referred to it (which was not often) they used geometrical phrases such as "the incomensurability of the side and the diagonal" of a square. (p. 48)

In light of Fowler's (1999) reconstruction of a pre-Euclidean mathematics based centrally on a notion of ratio (another meaning for the Greek word *logos*), as mentioned earlier *alogos* could simply mean "without ratio". Yet, Fowler went on to argue that if ratio did indeed refer to the sequence of *arithmoi* (his 'repetition numbers') generated by the process of *anthyphairesis* (continued subtraction in turn, 'the lesser from the greater' – what we think of as the Euclidean algorithm), then there *is* a ratio between the side and diameter of a square. Admittedly, it is an infinite sequence of *arithmoi,* but it is entirely predictable and provably regular.

Earlier mention was made of the *tetraktys* as an emblematic configuration for the Pythagoreans. However, other accounts suggest a different symbol playing a similar role for them, namely the pentagram (a regular pentagon with all diagonals drawn in). The side and diagonal of a regular pentagon are also incommensurable (indeed, they are always in the golden ratio) $\varnothing{:}1 = \frac{1+\sqrt{5}}{2}$. With regard to the process of *anthyphairesis* applied to the side and diagonal of the pentagon, we obtain the simplest infinite repeating pattern of *arithmoi* 'once, once, once, once, …' – simpler even than that between the side and diagonal of a square, which produces the sequence 'once, twice, twice, twice, twice, …'. (For more detail, see Fowler, 1999.)

Incommensurability does not entail indeterminacy. To say that the side and the diagonal of a square are incommensurable is, in part, to claim that neither may be expressed fully in units or perfectly equivalent parts of the other. But much can be ascertained about the relationship between the two. Notwithstanding Aristotle's complaint that the Pythagoreans failed to separate number from the things numbered, it seems wholly plausible to suggest that the Pythagoreans were not confused at all. Even the much-repeated tale of the expulsion and murder of the Pythagorean Hippasus for revealing this secret to the outside world and the concomitant Pythagorean 'crisis' of incommensurability could be a complete fabrication. [1]

The Meaning of Pattern

Within the context of profound possibilities that the Pythagoreans explored in patterns, it must count as one of the great ironies that the concept of 'pattern' itself is often viewed apologetically as a weak, catch-all notion unsuited for serious theoretical investigation. Even Gombrich, whose use of the concept has been shown in fact to be so skilled, felt compelled to acknowledge (in *The Sense of Order*) that he used it only by default, unsuccessful as he was in locating a more precise term with which to articulate his subtle points:

> There remains that jack-of-all-trades, the term 'pattern', which I shall use quite frequently though not with a very good conscience. For the word is derived from Latin *pater* (via patron), and was originally used for any example or model and then also for a matrix, mould or stencil. It has also become a jargon term for a type of precedent and has therefore lost any precise connotation it may have once had. (1979, p. x)

But is the absence of precise connotation a weakness or may it actually be a sign of strength? Gombrich quickly consoled himself with the following admission:

> Luckily it is a mistake to think that what cannot be defined cannot be discussed. If that were so we could talk neither about life nor about art. (p. x)

Of course, there are many other examples of such powerful words that defy precise definition – or better, that defy, or at least resist, what could be called *determinate* definition. To call words like 'art', 'life' or 'beauty' *vague* is not to demean, but rather to acknowledge that the concepts they mark out resist formulation in propositionally definitive lexical terms. Such determinate definitions, as a form, aspire to identify sets of necessary and sufficient conditions for any legitimate use of a term.

Often, however, especially in a living language, concepts not only connect in linear, logical ways but also enter into complex relationships with other, related notions, and metaphorically with yet others still. This is the case with many of the most important concepts, until they become quite thoroughly enmeshed in complex and interconnected patterns of human meaning and purpose.

Philosophers, following Ludwig Wittgenstein (1953/1963), sometimes refer to such concepts as 'open' and point to the notion of 'family resemblances' (developed by Wittgenstein in his *Philosophical Investigations*) as a way of showing how words may still have quite sound meanings, even if their meanings cannot be represented definitionally in neat, propositional terms. Wittgenstein illustrated this possibility with the notion of 'game'. One may discern a regularity in its uses, even though there is no common set of necessary and sufficient conditions to be discerned in its various applications. The concept underlying such words may in fact be a very complex one in respect of:

> [its] complicated network of similarities overlapping and criss-crossing: sometimes overall similarities, sometimes similarities of detail. (p. 32)

But this concept may also be quite potent with respect to the thematic threads that may be inferred from examples of its use:

> And the strength of the thread does not reside in the fact that some one fibre runs through its whole length, but in the overlapping of many fibres. (p. 32)

Just as a given pattern may be discerned as a pattern minus a determinate set of properties common to all its related examples, it is likewise the case that the concept of 'pattern' itself may be quite meaningful while missing a single set of necessary and sufficient defining conditions that capture all its legitimate uses. It is for this absence of a clear set of defining conditions that Gombrich worried about relying upon the concept of 'pattern' in his theoretical discussion. With the help of Wittgenstein, we can see why it is for precisely the same reason that Gombrich need not have be so concerned.

However difficult or impossible it might be to provide a determinate definition for 'pattern', it is still possible, again following Wittgenstein, to attempt to identify one or two of the more important themes or 'threads of similarity' that run through ranges of its use. Such an attempt is made in the next paragraph. It is important to note that no claim is made that in this characterisation the theme identified is at work in all legitimate uses of the term, nor that one may not use the concept of 'pattern' legitimately if it is absent. Notwithstanding these caveats, I do hold that the theme identified *is* consistent with all the uses of 'pattern' that this chapter has considered.

At least in so far as a provisional account of pattern is concerned, therefore, to discern a pattern is to see or consider something as part of an ordered arrangement such that it is possible to identify at least one of the principles constituting that order. To say an arrangement is ordered is to claim that the relationships among the arranged phenomena are not arbitrary, that the arrangement may be at least partially described in terms of one or more relational principles or themes.

In this respect, patterns need not be fully describable, their ordered arrangements need not be completely determinate. A pattern may be discerned across a number of phenomena (like a pattern in historical events) or within a given phenomenon (like the pattern in a specific work of art). Moreover, the determinate patterns pursued in most of the branches of mathematics in Devlin's account are reflected in this theme, as are the more open-textured applications of the concept in the work of Bateson and Gombrich. Bateson presented a challenge by characterising the pattern which connects as an unfixed pattern, "*primarily* (whatever that means) a dance of interacting parts". In an important sense, the theme identified above, in providing an accompaniment to this dance, can help complete Bateson's thought.

Finally, and most interestingly, this theme assists in locating a fresh epistemological perspective on the Pythagorean preoccupation with number as ordered plurality, one that may ultimately – as will be suggested in the remainder of this chapter – have enormous contemporary significance.

Pattern for Philolaus

Unlike 'order', which may often be considered in almost exclusively abstract and formal terms, there is something palpable about 'pattern' that reaches directly into the world of the senses and experience. The fact that the Pythagoreans most likely saw one of our more abstract contemporary concepts (number) in the positive terms of patterned entities tells us much about their view of mathematics and the world. It helps explain, for instance, their serious, even mystical, preoccupation with number – as ordered plurality – in the very fabric of the knowable world and their efforts to find the principles unifying that fabric in the mathematical–aesthetic ratios of harmonic intervals.

One of the most impressive views commonly attributed to these early philosophers, therefore, is the belief that the universe reveals itself to us in terms of number and principles of harmonious arrangement that are themselves expressible mathematically. Put that way, the Pythagorean position would appear a paradigmatically modern view, although it is customary to trace back the provenance of the modern view respecting the efficacy of mathematics in describing physical reality to Galileo's imagistic claim that the universe is a book written in the language of mathematics.

Galileo's observation does indeed mark a turning point in the development of a genuinely scientific method, but perhaps more credit should be given to the Pythagoreans in this regard. Certainly their work proceeded within a framework of metaphysical presumptions respecting the ultimate nature of reality, one that we now may view as quasi-religious or mystical.

Nevertheless, many contemporary philosophical investigations into the foundations of scientific method have shown that all science proceeds on certain metaphysical presumptions that lie outside the empirical reassurances of either verifiability or falsifiability.

What can be said, what can be shown

Moreover, it is likely that the ultimate reality of the Pythagoreans was viewed by them as *unknowable in principle* by human beings. A very modern view itself, this perspective is much in the spirit of Wittgenstein's (1922/1958) proclamation at the very end of the *Tractatus Logico-Philosophicus:* "Whereof one cannot speak, thereof one must be silent" (p. 189). Wittgenstein held his own metaphysical achievement in that work to be a heuristic trope, a figurative 'ladder' that could be discarded once it had done its work:

> My propositions are elucidatory in this way: he who understands me finally recognizes them as senseless, when he has climbed out through them, on them, over them. (He must so to speak throw away the ladder, after he has climbed up on it.) He must surmount these propositions; then he sees the world rightly. (p. 189)

In Wittgenstein's logically austere foundations of science, therefore:

> There is indeed the inexpressible. This *shows* itself; it is the mystical.
> (p. 187; *italics in original*)

There is indeed the supra-empiricist and supra-conceptual metaphysical framework that shows itself figuratively, indeed poetically, through sets of admittedly 'senseless' propositions. In the *Tractatus,* as in that other masterpiece of modernist philosophy of science by A. J. Ayer (1952), *Language, Truth and Logic,* the inexpressible insights enabling scientific inquiry and other non-empirical or non-tautological insights could be shown figuratively and apprehended viscerally, even though they could not be articulated in propositional form.

As a final, modern touch, we shall soon see that the claims made by the Pythagoreans for the pervasiveness of 'number' in the universe – as knowable by human beings – may plausibly be taken to relate to the positive, human reality of experience and perception. They may, in fact, be intended to relate to discoverable, empirical patterns.

The Pythagoreans and Philolaus

Within a largely a-textual tradition, there is one piece that has come down to us in fragments – the first written by a genuine Pythagorean, one Philolaus of Croton. It provides tantalising and vivid glimpses into the nature of these Pythagorean views and the crucial role the concept of pattern plays therein. In light of the suggestion just offered that Philolaus and the Pythagoreans anticipated the modern view of Galileo on the role of mathematics in understanding the universe and other modern perspectives, it is more than interesting to note that Copernicus himself noted that the "stimulus for his revolutionary cosmological system" came from ancient sources and that "in this connection he twice names Philolaus" (Burkert, 1972, p. 337). Indeed, at the time of Copernicus, and for a good time thereafter, the Copernican system itself was referred to as *astronomia Philolaica* or *astronomia Pythagorica.*

It is not with these most prescient cosmological views, however, that this chapter is concerned, but rather with two short fragments (numbers 4 and 6) that turn on the nature and possibility of human knowledge. In considering these Pythagorean fragments, some interpretations of Philolaus and the Pythagorean Order will be offered.

Philolaus of Croton

Of Philolaus, Walter Burkert wrote:

> Practically nothing is known of his life. His home was Croton, or maybe Tarentum, and he spent some time in Thebes – all the rest, what little there is of it, is demonstrable embellishment or simple misunderstanding. (1972, p. 228)

To this dearth of factual knowledge surrounding the historical Philolaus may be added the traditional problems associated with early Greek thinking. Such fragmentary textual evidence as we do have often comes down to us tendentiously, cited by other writers with their own powerful philosophical presumptions to propound or defend. In addition to this textual morass, there also are long, minutely detailed and intellectually impressive debates among scholars concerning the authenticity as well as the interpretation of each part of the fragmentary record. Such debates are well beside the point of this exploratory effort, except to assist in noting that in a textual environment that admits of so many possible interpretations, any plausible interpretation may usefully find a place.

Carl Huffman (1993) places Philolaus solidly within the context of pre-Socratic philosophy. The Ionian philosophers Thales, Anaximander and Araximenes, those famed sixth-century originators of systematic, speculative inquiry into the nature and origins of the physical world, sought to identify the basic principles at work in the world in a properly philosophical way. They pursued their inquiries independently of any religious or mythopoeic explanations and developed their positions according to defensible rational standards.

Each of these three posited an underlying 'stuff' out of which all worlds must arise. For Thales, this primary material was held to be water; for Araximenes, it was air; for Anaximander, it was boundless, indeterminate *apeiron* from which the elements are formed, but different from any of them Opposition, change and transformation become the sustaining principles within worlds as systems, with things as they appear to us in this world being simply mutable expressions of more primary materials. To the mutable physical reality of their cosmology, the Ionians also challenged the religious and cultural ethnocentrism of the Greeks. Xenophanes noted sharply the ways in which different races pictured gods in their own distinctive images and, even more tellingly, the manner in which projected an anthropomorphic structure onto their identity.

Philolaus was, however, agnostic regarding these sorts of philosophical conclusions. While his cosmology included a central fire around which the earth orbits, Philolaus made no claims for fire as the universal *Urstoff*. What Philolaus offered epistemologically was, in a remarkably modern vein, a view on the *limitations* of human knowledge. The core of fragment number 6, in Huffman's definitive 1993 translation, presumes that:

> Concerning nature and harmony the situation is this: the being of things, which is eternal, and nature in itself admit of divine and not human knowledge. (p. 123)

More anticipations of the modern?

Nature in itself – the face of reality as it would be configured in the mind of God – is unknowable to human beings. What Kant, more than two thousand years later, would call the *noumenal* world, the world of things-in-themselves, was, for Philolaus, as it would be for Kant, not susceptible to human representation. What humans do know, what it is *possible* for us to know – for Kant – are the phenomena as structured by the cognitive 'scaffolding' comprising the bounds of human sense, structures of reason involving those *a priori* concepts immanent in the human mind itself. Given the ways in which modern ideas seem to be prefigured in the thinking of Philolaus and the Pythagoreans, one might well wonder if some early comparable version of the Kantian scaffolding was at work among them as well. The speculation that this chapter is exploring is affirmative in that regard and offers – as a plausible, provisional hypothesis only – that the pseudo-Kantian scaffolding in Philolaus involves the concept of number or ordered plurality.

The importance of any cognitive scaffolding theory in our time of rapidly expanding interest in embodied cognition is manifest (see, for example, Lakoff and Nuñez, 2000). In these contemporary terms, in which the locus of mind increasingly is to be discerned in the viscera and physiology of human beings, we should have no difficulty in understanding a pseudo-Kantian position on intellectual scaffolding. Unlike Kant, who viewed it as a disembodied feature of pure reason, we might more comfortably express it scientifically in terms of genetically achieved physiological and especially neuro-physiological predispositions that undergird our capacity for perception and language.

In an embodied cognitive model, this scaffolding would be seen to support our ability to discern and represent phenomena in ways our specific evolutionary history has found advantageous. Two key ingredients in any such embodied predisposition must surely be captured in Kant's synthetic *a priori* position on causality. Human knowledge and action proceed on the principle that every event has a cause, a form of efficacious connective tissue. However, those causal links cannot, as Hume, Kant and many other philosophers have maintained, be empirically discovered, observed or conceptually deduced, apart from a theoretical framework whose explanatory *concepts* provide whatever causal efficacy may be held to do the linking.

As the eminent philosopher of science Norwood Hanson wrote in 1958:

> Causes certainly are connected with effects; but this is because our theories connect them, not because the world is held together by cosmic glue. (p. 64)

In the properly agnostic vein of Kant and Philolaus, Hanson went on to observe:

> the world *may* be glued together by imponderables, but that is irrelevant for understanding causal explanation. (p. 64; *italics in original*)

An embodied presumption of caused events is likewise not so difficult to understand in biological terms. It is difficult to imagine the human species or its many predecessor species having much viability without a biological predisposition to discriminate among events *as* possible effects and to discriminate among specifiable phenomena *as* potential causes. Hence, the propensity to represent the world in terms of discrete entities linked cognitively and meaningfully by organising concepts would likewise appear to be an important part of our evolutionary strategy. In these terms, therefore, one might well be tempted to see enormous contemporary significance in Philolaus's fourth fragment:

> And indeed all the things that are known have number. For it is not possible that anything whatsoever be understood or known without this. (in Huffman, 1993, p. 172)

Number as Cognitive Scaffolding

For Philolaus, therefore, everything that is knowable has number. Recalling that by number Philolaus meant *ordered plurality,* this means in one important sense that everything knowable is *patterned*. In order to be able to know, we need to be able to *identify* a range of phenomena as discrete (i.e. denumerable) phenomena of a certain kind. Regardless of whether or not reality is ultimately one integral 'stuff' in the mind of God, in order for human beings to achieve any efficacy in our environments, we need to be able to distinguish individual phenomena of one or another kind.

These phenomena, better, these *sets* of phenomena, therefore have number. To know them is to discern the pattern at work among them, that is to understand the principle(s) by which they are ordered. Similarly, to have a concept is to have discerned a pattern in the uses of an expression.

Hence, everything known or knowable must have number, must exhibit an ordered plurality. And so for Philolaus, the mathematical – in that special early Greek sense in which the mathematical and the aesthetic seem to commingle so productively – is at work in all human knowledge. As a final speculation, is it plausible to suggest that in the Pythagoreans' views on number and knowledge, we may discern a fundamental commonality between the mathematical and the aesthetic of enormous potency? Might this commonality have been for them the cognitive infrastructure or 'scaffolding' for knowing?

Surely, there is a *special* sub-pattern to be discerned synoptically in mathematics and art. If mathematics may productively be viewed as the creative exploration and formal representation of pattern possibilities, art may equally and symmetrically be viewed as the creative exploration and sensory representation of pattern possibilities. Are we dealing with two 'subjects' or 'disciplines' here, like history and geography, or with distinctive expressions of a common patterning proclivity, scaffolding or cognitive infrastructure? Is it this common patterning proclivity that enables us to discern and to make patterns, either physically or symbolically?

To view knowledge as pattern on this kind of account is to view history, geography, science, etc. as the pursuit of discoverable patterns in physical, social and symbolic environments and their representation, with number as ordered plurality as its epistemological engine. Without the enabling infrastructure of number in the profound sense of Philolaus and the Pythagoreans, would we be able to *have* a history or a geography or a science? Remove our ability to discern pattern and would we be able to *know* anything?

Note

[1] The claim 'The number $\sqrt{2}$ is irrational' would seem to imply that the notion, indeed the phenomenon, of irrationality (or the linked geometric notion of incommensurability) was pre-existent and that this result were simply a matter of showing this fact for the particular instance of $\sqrt{2}$. However, after providing a proof of this result in terms of the side-to-diagonal incommensurability of the square (involving *reductio ad absurdum* arguments about particular numbers of units necessarily being both odd and even) – and bearing in mind that, according to Aristotle, investigation of 'the odd and the even was of particularly fundamental significance to the Pythagoreans' – the historian of mathematics John Fauvel (1987) argued:

> It is not known whether this was the original proof of the result that the side and the diagonal of the square are incommensurable. Nor is it known whether the case of the square was the first in which the phenomenon of incommensurability was recognised. (Some historians have argued that this recognition took place in connection with investigations of the regular pentagon, whose side and diagonal are also incommensurable.) It is not important to resolve this for the purposes of our story. But notice one significant aspect of the result you have just seen proved: assuming you

found it convincing, and now believe the result, you do so *only* because of the *proof.* The result has very little plausibility without proof accompanying it. This is an entirely new situation. Other results earlier in the [undergraduate mathematics student] unit – Hippocrates' quadrature of lunes, say, or the unlimited number of primes – had proofs which acted more so as to corroborate what might have seemed quite likely beforehand. But the discovery that two lines were incommensurable, and the proof, must have been more-or-less simultaneous. Indeed we might go further and say that its first proof must have constituted its discovery, though the details of this event are no longer known. (p. 18; *italics in original*)

Admittedly, the examples Fauvel gives of Hippocrates and certainly Euclid date from at least a century later than the first Pythagoreans. But his final observation seems most telling: in an important sense, the very phenomenon, the very concept of incommensurability must (to use Lakatos's term) have been a proof-generated one. (See also Fowler, 1993.)

CHAPTER 6
Mathematics, Aesthetics and Being Human

William Higginson

> The problem is more aesthetic than ethical, philosophical, sexual,
> psychological, or political, though it goes without saying that such
> divisions are unacceptable to me because *everything* that matters
> is, in the long run, aesthetic. (Mario Vargas Llosa, 1999, p. 194)

The general intellectual terrain in which this chapter is situated is large,
heavily trafficked and contentious. The underlying scholarly question that
circumscribes it – 'What does it mean to be human?' – has been actively pursued since classical times by both humanists and scientists. In addressing
this issue, philosophers have often connected it to questions of cognition.
For instance, in the opening paragraphs of his book *On Human Nature,*
Harvard biologist E. O. Wilson (1978) wrote:

> These are the central questions that the great philosopher David
> Hume said are of unspeakable importance: How does the mind
> work, and beyond that why does it work in such a way and not
> another, and from these two considerations together, what is man's
> ultimate nature. (p. 1)

A great deal of the extensive discourse around this question in the past three
decades has been driven by scientific and technological advances in the biological sciences. Wilson has been just one of many scientists to make the case
for socio-biological underpinnings of much human behavior. Social scientists
have not been slow to pursue the implications of some level of genetic predisposition. Consider, for example, Charles Murray (2003), in the introduction
to his book *Human Accomplishment,* who draws attention to two core human
impulses:

> The first is the abiding impulse of human beings to understand, to
> seek out the inner truth of things. [...] The other impulse is *Homo
> sapiens'* abiding attraction to beauty. [...] Many of the most endur-
> ing human accomplishments have been, simply, things of beauty.
> (pp. xix-xx)

Murray's choice of what he considers to be the two most prominent 'embedded' characteristics of human beings fits the theme of this chapter particularly well. This is because I am especially interested in the ways in which
mathematics – seen conjointly as an artifact and an activity – both forms,
and is formed by, human abilities and cultures. [1]

Other contributions to this book have included many compelling demon-strations of the role of the beautiful in mathematical functioning and varied and articulate views of mathematicians themselves about this issue. Some-what by contrast, in this chapter I want to move outside of this particular 'insider' arena in two different directions.

For my first shift, I look in more detail at what the world at large (at least the world as portrayed in more popular and populist culture, as well as the images of schoolchildren) believes to be the case about mathematics and mathematicians. There is, of course, something uncomfortably familiar about the baleful looks perennially cast at the subject, its institutional purveyors and, most certainly, its perpetrators. However, I wish to look at how this public image might plausibly be argued to be undergoing something of a sea change.

Then, in a second move, I begin to explore what certain humanist fig-ures in the arts and social sciences have had to say about the centrality and importance of mathematics to human concerns, creativity and awareness. Finally, at the end of my chapter, I return to examine in more depth my pro-posal of the possibly essential mathematical character of human beings.

But first, I foreshadow this later discussion by a brief excerpt from a set of lectures by the eminent literary critic George Steiner, who, in 1990, deliv-ered the Gifford lectures at the University of Edinburgh on the idea of cre-ation in Western thought, literature, religion and history. When published in significantly elaborated form more than a decade later (Steiner, 2001), even readers familiar with Steiner's eclecticism and penchant for academic pere-grination were startled, and in the case of some reviewers befuddled and annoyed, by the central role given to mathematics in his consideration of the wellsprings of human creativity.

> To Plato, the point would have been self-evident. It is inconceiv-able that one should question a life of the mind without address-ing mathematics and the sciences which, in the main, derive from the sovereignty of mathematics. Since Galileo and Descartes, this injunction has become theoretically and pragmatically inescapable. It is in mathematics and the sciences that the concepts of creation and of invention, of intuition and of discovery, exhibit their most immediate, visible force. […]

> The difficulty, however, is twofold. Mathematicians and scientists "get on with the job". […] they avoid too close a scrutiny of the epistemological foundations of their disciplines. […] The second difficulty is one of access. […] One needs considerable familiarity with mathematical symbolism in order to follow the controversies on whether or not there are in pure mathematics "discoveries" or, instead, an autonomous unfolding of *a priori*, as it were tautological, systems generated from within the human intellect and its deep-seated instinct for speculative, other-worldly play. *Homo ludens*. If, as Galileo ruled, nature speaks mathematics, far too many of us remain deaf. […]

> It is at this subliminal level that decisive choices are arrived at as
> between a congeries of possible, though rule-bound, combina-
> tions. [...] But how does the sub-conscious choose? [...] What is
> arresting is the move towards the aesthetic. [...] The "useful com-
> binations," where "useful" signifies the generative strength which
> will lead to further propositions, to related theorems and general
> laws, "are precisely the most beautiful". (pp. 176-178)

In drawing these initial remarks to a close, and in the spirit of the opening
quotation from Vargas Llosa, I choose to interpret Murray's two selected
'aspects of human nature' as two variations on an aesthetic theme. The drive
to understand and the attraction to beauty can both be seen as manifestations of a universal human ability to sense what 'fits' in a given situation and
what does not.

I wish to take this assertion two steps further and then add a necessary
caveat. First, I want to contend that the roots of all mathematical activity are
located close to this aesthetic predisposition. Second, I claim that this argument parallels in many important ways the sustained argument made by Ellen
Dissanayake (1995) that humans are inherently aesthetic in their approach to
the world. I wish to acknowledge here the influence of her speculations about
Homo Aestheticus (the title of her book) on my vision of humans as inherently
mathematical beings.

However, I also need to acknowledge that much in the world, especially
the pervasive and resilient public image of mathematics and, alas, mathematicians, seems set in quite the opposite direction. So it is in this latter context that I start my explorations here, with a formative recollection that
served to trigger my own interest in this area.

Atlantic Primes

Scholarly interest in particular topics can sometimes be connected to particular events. The roots of this chapter go back to an experience I had more
than twenty years ago, while looking after two five-year-olds; my daughter,
Kate, and Aaron (not his real name), the son of friends. As perhaps befitted
healthy children of this age, with educated and supportive parents, the two
youngsters were energetic, intellectually curious and articulate. In addition
to these general characteristics, Aaron had exceptionally advanced capabilities in both mathematics and language. Numbers fascinated him and he was
already – as a pre-schooler – an avid reader. Knowing this, I decided to
make him aware of something I had just read.

The piece in question had a rather complex history. Appearing in the
April 1980 issue of the venerable American periodical *The Atlantic Monthly,* it
had been written by Horace Judson, a science writer whose book on the evolution of the field of molecular biology, *The Eighth Day of Creation* (1979),
had received exceptionally positive reviews. As part of his research for *The
Eighth Day,* Judson had interviewed a large number of leading scientists

about what motivated them and the satisfactions they received from their work. Intrigued by what they had said, he wrote, as something of an off-shoot of his main work, an introductory book on the philosophy of science, called *The Search for Solutions* (1980a). In an attempt to capture, early in his publication, something of the passion that appears to be a universal motivator of scientists, Judson began Chapter One, entitled *Investigation: the Rage to Know,* with a story about one of his friends.

> Certain moments of the mind have a special quality of well-being. A mathematician friend of mine remarked the other day that his daughter, aged eight, had just stumbled without his teaching onto the fact that some numbers are prime numbers [...] "She called them 'unfair' numbers," he said. "And when I asked her why they were unfair, she told me, 'Because there's no way to share them out evenly'." What delighted him most was not her charming turn of phrase nor her equitable turn of mind [...] but – as a mathematician – the knowledge that the child had experienced a moment of pure scientific perception. She had discovered for herself something of the way things are. (p. 2)

It was this first chapter of *The Search for Solutions* that was published as an article in *The Atlantic* under the title 'The rage to know' (1980b). Given that he was writing for lay people, Judson had felt compelled to explain just what prime numbers are. Hence, part-way through the passage noted above, we read, "prime numbers – those like 11 or 19 or 83 or 1,023, that cannot be divided by any other integer [except, trivially, by 1]". Judson was a highly intelligent, well-educated individual. Despite this, in his attempts to clarify, he made an error in one part of his statement. This was the observation I decided to share with Aaron.

"I read something interesting in *The Atlantic* today, Aaron", I said. "A man wrote that 1,023 was a prime number." Aaron's response was immediate and definite. With eyes wide and his face somehow managing to express both delight and dismay, he responded, "Oh, no! 1,023 is 3 times 11 times 31." Shortly thereafter I heard him say excitedly to Kate, "Some man said in the ocean that 1,023 was prime!"

I never did pursue just what my daughter made of that particular observation. I was conscious, however, that this incident had planted a seed for me. How could it come to be that a species might exhibit this exceptional range of behavior with many highly educated adults failing to match the insights of a very young, formally untutored, child? Just what was this domain where such things happened? What were some implications of this? Under what circumstances might they be altered?

There are, of course, orthodox answers to these questions, most of them pointing, lazily, in the direction of inexplicable genius and an extreme form of élitism. According to this view, mathematics, as has been mentioned in earlier chapters, is seen as the prerogative of a special few – a few who often pay a high human price for their powers of abstraction. [2] Lurking not very far beneath the surface are parallels with the magical and wizardly. The subject is useful, but potentially dangerous. Societies very much need mathematical insights, but as individuals we might perhaps be relieved to be unburdened by the gift of mathematical precocity.

Some Possibly Less-Familiar Examples of a Familiar Stereotype

I attempt here to illustrate ways in which some of the different elements of the orthodox stereotype of mathematicians (and, indirectly, of mathematics) have been reflected and perpetuated in the writings and actions of a number of influential thinkers over the past two centuries. Stereotypes are, by their nature, not concepts that invite critical analysis and this one is no exception.

In cases where some of these issues have been the focus of reflection, there has been a tendency to recycle a small number of archetypical examples (much as some of the mathematical examples of the aesethetic have been). Hence, as the example of mathematical precocity we have frequent repetition of the story of the schoolboy Gauss shocking his teacher with his rapid calculation of the sum of the first one hundred natural numbers. The equivalent story at the level of the mathematical researcher and the question of mathematical creativity most likely is the account by Poincaré of his work on Fuchsian functions.

To omit these examples is not to deny their validity, although my sense is that the Gauss example is not as strong as it is often assumed to be. I have attempted to dig a little deeper and to find some lesser-known examples that come from a more extended range of individuals. In the views I shall provide, articulated by Charles Darwin, Alfred North Whitehead and C. P. Snow, there are intimations of power, pervasiveness, precocity and a whiff of both the divine and the delicious, and the not completely rational. There is a distinct sense of a pecking order, of the learning experience as being crucial to later attitudes and a strong stereotype of the mathematician as social (at times awkwardly so) eccentric.

Darwin, Whitehead and Snow

The first set of images comes from Charles Darwin, who, when in his mid-sixties, wrote, with his children as his intended audience, an informal biography (1887/1958). In the section where he recalls his undergraduate years, Darwin noted:

> During the three years which I spent at Cambridge my time was wasted, as far as the academical studies were concerned, as completely as at Edinburgh and at school. I attempted mathematics, and even went during the summer of 1828 with a private tutor to Barmouth, but I got on very slowly. The work was repugnant to

> me, chiefly from my not being able to see any meaning in the early
> steps in algebra. This impatience was very foolish, and in after years
> I have deeply regretted that I did not proceed far enough at least
> to understand something of the great leading principles of mathe-
> matics, for men thus endowed seem to have an extra sense. (p. 18)

This passage seems to invoke a very similar idea to that expressed by George
Steiner a hundred and fifty years later. In a subsequent passage, Darwin's son
Francis, who served as editor for his father's autobiography, noted:

> My father's letters to Fox show how sorely oppressed he felt by the
> reading for an examination. His despair over mathematics must
> have been profound, when he expresses a hope that Fox's silence
> is due to "your being ten fathoms deep in the Mathematics; and if
> you are, God help you, for so am I, only with this difference, I
> stick fast in the mud at the bottom, and there I shall remain". Mr.
> Herbert says: "He had, I imagine, no natural turn for mathematics,
> and he gave up his mathematical reading before he had mastered
> the first part of algebra, having had a special quarrel with Surds
> and the Binomial Theorem." (p. 114)

Some five decades after Darwin's reflections, the Anglo-American mathematician and philosopher Alfred North Whitehead delivered the Lowell lectures at Harvard University. These lectures which the author saw as "a study
of some aspects of Western culture during the past three centuries in so far
as it has been influenced by the development of science" (p. ix) constituted
the core of his *Science and the Modern World* (1926). The second chapter
of this book is called 'Mathematics as an element in the history of thought'
and in it we find the following observation:

> The science of Pure Mathematics, in its modern developments,
> may claim to be the most original creation of the human spirit. [...]
> There is an erroneous literary tradition which represents the love
> of mathematics as a monomania confined to a few eccentrics in
> each generation [...] Even now there is a very wavering grasp of
> the true position of mathematics as an element in the history of
> thought. I will not go so far as to say that to construct a history of
> thought without profound study of the mathematical ideas of suc-
> cessive epochs is like omitting Hamlet from the play which is
> named after him. That would be claiming too much. But it is cer-
> tainly analogous to cutting out the part of Ophelia. This simile is
> singularly exact. For Ophelia is quite essential to the play, she is
> very charming – and a little mad. Let us grant that the pursuit of
> mathematics is a divine madness of the human spirit, a refuge from
> the goading urgency of contingent happenings. (pp. 29-31)

Once again, George Steiner, seventy years on, would seem to suggest that
omitting mathematics now would be more like losing Hamlet himself.

   For the third of these vignettes, I turn C. P. Snow's 1959 Rede Lecture,
entitled *The Two Cultures and the Scientific Revolution*. The theme of the

lecture was that a deep and widening gulf existed between the worlds of scientists and humanists. As part of his argument, early in is lecture, Snow recounted a 'high table' story:

> By this I intend something serious. I am not thinking of the pleas-
> ant story of how one of the more convivial Oxford greats dons –
> I have heard the story attributed to A. L. Smith – came over to
> Cambridge to dine. The date is perhaps the 1890's. I think it must
> have been at St. John's, or possibly Trinity. Anyway, Smith was sit-
> ting at the right hand of the President – or Vice-Master – and he
> was a man who liked to include all round him in the conversation,
> although he was not immediately encouraged by the expressions
> of his neighbours. He addressed some cheerful Oxonian chit-chat
> at the one opposite him, and got a grunt. He then tried the man
> on his own right hand and got another grunt. Then, rather to his
> surprise, one looked at the other and said, 'Do you know what
> he's talking about?' 'I haven't the least idea.' At this, even Smith
> was getting out of his depth. But the President, acting as a social
> emollient, put him at his ease, by saying, 'Oh, those are mathemati-
> cians! We never talk to *them*. (p. 3; *italics in original*)

Just why Snow should think that this is a "pleasant story" is not entirely clear. For my purposes, interested in the way mathematics and mathematicians are perceived in society, it is significant to note that this is one of the few places in the lecture, or in the substantial related material that Snow wrote in subse-quent years, where he uses examples from other than the physical sciences.

Peirce *père et fils*

The figure of Benjamin Peirce cuts a wide swath in the history of American mathematics. Frequently identified as the first major mathematical thinker to reside in North America, he was the author, in 1870, of *Linear Associative Algebra,* a text where he presented mathematics as "the science which draws necessary conclusions". Born in 1809, he joined the Harvard faculty in the early 1830s and remained there for almost five decades. He is best known in some intellectual circles as the father of the brilliant but erratic philosopher and mathematician, Charles Sanders Peirce.

The Peirces, father and son, came to the attention of a wider than usual audience in 2001 when Louis Menand included Charles as one of the four key thinkers (along with Oliver Wendell Holmes, William James and John Dewey) in his Pulitzer Prize winning book, *The Metaphysical Club*. Menand's portrait of Benjamin Peirce begins:

> Peirce was probably the first world-class – in the sense of interna-
> tionally recognized – mathematician the United States produced.
> He cultivated a certain wizardliness of manner. His hair was iron-
> gray, and he wore it long, with, in later years, a thick beard. And
> his obscurity was legendary. It was said at Harvard that you never
> realized how truly incapable you were of understanding a scientific

matter until Professor Peirce had elucidated it for you. [...] Peirce enjoyed the reputation and even played up to it, because he was a confirmed intellectual elitist, a pure meritocrat with no democracy about him. "Do you follow me?" he is supposed to have asked one of his advanced classes during a lecture. No one did. "I'm not surprised," he said. "I know of only three persons who could." [...] It was Peirce's view that mathematics was the supreme science, but a science accessible only to a few. (p. 153)

Menand's image of Peirce senior is supported by several passages in Joseph Brent's (1998) biography of Charles Sanders Peirce. There we find:

Charles's father was also by all accounts, a most unusual and unconventional man. Students remembered him with great respect and affection and thought of him as a genius who, as often as not, they were unable to understand. This description of a typical class was written by one of his students.

I have hinted that his lectures were not easy to follow. They were never carefully prepared. The work with which he rapidly covered the blackboard was very illegible, marred with frequent erasures, and not infrequent mistakes (he worked too fast for accuracy). He was always ready to digress from the straight path and explore some sidetrack that had suddenly attracted his attention, but which was likely to have led nowhere when the college bell announced the close of the hour and we filed out, leaving him abstractedly staring at his work, still with chalk and eraser in his hands, entirely oblivious of his departing class. (p. 32)

Shortly thereafter Brent notes that, "Benjamin Peirce taught mathematics as a kind of Pythagorean prayer" (p. 33). Many of his students saw him as "a real live genius, who had a touch of the prophet in his make-up" (p. 32). [3]

I have quoted Menand and Brent at length because the picture they paint of Benjamin Peirce resonates in many ways with descriptions of mathematical researchers and teachers found in other books and papers. Take, for instance, the well-known image of the "traditional mathematics professor" (Pólya, 1957, p. 208) who "writes a, he [sic] says b, he means c, but it should be d". Or, the less amusing but real description of a young mathematics professor whose:

lectures were useless and right from the book [...] He showed no concern for the students [...] He absolutely refuses to answer questions by completely ignoring the students. (Gibbs *et al.,* 1996, p. 41)

The instructional context for this comment was another leading mathematical research center, in this case the Berkeley campus of the University of California. The teacher was Ted Kaczynski, at that time a highly regarded early-career researcher, who was, quite a number of years later, to achieve great notoriety as the Unabomber.

The particular role played by Peirce in the establishment and perpetuation of the image of the mathematician as a brilliant but distanced and 'other worldly' individual is certainly debatable. But there is certainly no doubt that he fitted it. What seems much more certain is the robustness of the image of the mathematician in society.

Schoolchildren's images of mathematics and mathematicians

In the late 1960s, Cambridge psychologist Liam Hudson (1970) carried out a widely reported piece of research (published under the title *Frames of Mind*) where he examined the attitudes and perceptions of selected groups of male secondary school students to different professions. Using questionnaires and a 'semantic differential' technique, subjects rated a number of "typical figures (in this case, 'Novelist', 'Historian', 'Mathematician', 'Physicist', etc.) against pairs of adjectives ('warm/cold', 'intelligent/stupid', 'hard/soft', 'valuable/worthless', and so on)" (p. 46).

Hudson found a high degree of consensus in the responses. He reported:

> one is struck by the tendency for certain typical figures to cluster together: Mathematician, Physicist, and Engineer clearly resemble each other closely in schoolboys' minds; and so too do Poet, Artist, and Novelist. Examined in detail, this 'scientific' cluster can be broken down a little. The Engineer is seen as less intelligent, cold and dull than the other two, but as more manly, and dependable and imaginative. The Mathematician is seen as even colder, duller and less imaginative than the Physicist. At the other extreme, the Poet is seen as more intelligent than the Artist, but less warm and exciting; the Novelist is more imaginative than the Artist, but less soft and feminine. [...] If these two clusters – the scientific and the artistic – are set up as polar opposites, the other professions represented form a spectrum between them. (pp. 48-49)

Thirty years after *Frames of Mind* was published, an American and a British researcher (Picker and Berry, 2000) carried out an international study that replicated several aspects of the mathematical component of Hudson's research. Most of the negative elements of the stereotypical mathematician found in Hudson's work remained firmly in place in the minds of some five hundred lower secondary school students in the United States, the United Kingdom, Finland, Sweden and Romania. Part of their data comprised sketches the students made. The most common figure sketched by the students was an unkempt, glasses-wearing, balding, middle-aged white male.

The researchers went on to note that a "completely unexpected theme that emerged from the drawings" was that of "mathematics as coercion" (p. 74). In a significant percentage of the drawings pupils drew mathematicians as teachers who used intimidation, violence or threats of violence to 'make' their charges learn. The connection between mathematics and violence is returned to in the final chapter.

In the next section, I turn to examine more recent examples of portrayals of the nature of mathematics and its human connections and manifestations. Many of these examples are generated, at least indirectly, by variations on the theme of embodiment emerging from the biological and cognitive sciences. Others are to be found in a new location, the arts. The last decade has seen a rash of books, movies and plays with broad and non-trivial mathematical content (Emmer and Manaresi, 2003). On the surface, many of these seem only to perpetuate the standard view of the creative mathematician as the eccentric genius. But by looking more broadly and deeply, I suggest one can perceive a shift in the direction of seeing mathematics as a natural human ability. This is not to deny the existence of a broad range of that ability, but it does challenge the 'zero/one' model of the nature of mathematical ability, which sees it as the exclusive preserve of a small, predominantly male, élite.

Mathematics and the Human, Revisited

In very broad terms, there has been the empirical 'rejuvenation' of mathematics from two different but related directions. More in keeping with the field's traditional alliances with the 'hard' sciences, there have been influences emerging from the rapid and highly mathematised development of the various instruments of new information technology. Co-ordinated by ever more powerful computational tools, several fields, including mathematics, have entered into a period of considerable transformation. As one case in point, the development of 'experimental' and 'constructive' approaches to the discipline have appeared that would have been inconceivable to most scholars in the field even twenty years ago (Borwein and Bailey, 2003).

Technology has also been a factor in the links being forged from another direction, namely the biological and human sciences. The development of tools and techniques to probe 'life' at the molecular level have prompted yet another round of debates on human nature. In discussions that have perhaps capitulated far too much to the polarised, Aristotelian, either–ors of 'nature' and 'nurture', there has been a vigorous counter-attack on the part of scholars leaning toward the deterministic. Building in many cases on the broad groundwork laid by E. O. Wilson, and stated most aggressively by Stephen Pinker (2002) in *The Blank Slate: the Modern Denial of Human Nature,* there has been a reconsideration from many perspectives of the possibility that humans are 'hard-wired' to behave in certain ways.

These reconsiderations manifest themselves in many areas. The field of cognitive science, for instance, owes its relatively recent existence to the co-ordinated study of issues which in more classical times would have been relegated to specialised and isolated investigation by psychologists, philosophers, physiologists, logicians, anthropologists and even computer scientists (Gardner, 1987). The talented neurologist and gifted writer Oliver Sacks has been particularly successful in making interested readers more aware of the subtleties and power of the relationship between mind and body.

Approaches to intellectual questions that give primacy to the role of the physical are sometimes considered collectively under the term 'embodiment' (Weiss and Haber, 1999). One of the most active and influential thinkers in this area is the Berkeley linguist and philosopher, George Lakoff. His earlier work focused on linguistic aspects of embodiment, before moving, with one colleague in *Philosophy in the Flesh* (1999), to consider some philosophical connections of this theme. Shortly thereafter, with the assistance of another colleague, he carried his theories into the field of mathematics with *Where Mathematics Comes from: How the Embodied Mind Brings Mathematics into Being* (2000).

This text could be seen, from some perspectives, as the fourth venture into the terrain of 'embodied mathematics', being preceded by the publications of Dehaene (1997), Butterworth (1999) and Devlin (2000). These texts differ among and between each other and, while all have merit, it seems unlikely that they will, in the longer term, be seen to have resolved any major issues. What they did do, however, was to cause at least some members of the mathematical community to think differently about their own discipline.

Popularising mathematics and mathematicians

Over the same time period, and for some of the same reasons, the intellectual world has witnessed a dramatic shift in the availability of 'popular' treatments of two areas traditionally seen as being abstruse and difficult, namely aesthetics and mathematics.

In the traditional division of philosophy into its sub-disciplines such as ethics, epistemology, metaphysics and logic, aesthetics has been something of a 'Cinderella' case. Partly this has been a function of the 'bloodless' approach taken to issues in the area by many classical scholars. The 'new aestheticians' come, increasingly, from outside philosophy, narrowly defined. One direction of approach is from literature (Donoghue, 2003; Fisher, 1998; Scarry, 1999), another is from science (Fischer, 1999; McAllister, 1996; Wechsler, 1978), while a third arrives from environmental and architectural studies (Carlson, 2000; Hildebrand, 1999) and a fourth from the world of commerce (Postrel, 2003).

'Popular' treatments of mathematical ideas are not new. Authors like W. W. Sawyer, Lancelot Hogben, W. W. Rouse-Ball and Martin Gardner, for instance, wrote prolifically for large and appreciative audiences throughout the previous century. Nor is it unusual to have first-class researchers addressing the relationship between aesthetics and mathematics. George Birkhoff's (1933) *Aesthetic Measure* and Herman Weyl's (1952) *Symmetry* are particularly interesting exemplars in this area. That having been said, the proliferation of mathematically related books, plays, novels and movies in the past decade and a half, is, in its scale alone, without precedent.

It seems clear that the success of James Gleick's (1987) *Chaos* and Simon Singh's (1998) book *Fermat's Last Theorem* (and the related video) have been highly influential. But, for whatever reasons, the former trickle of publications has become a torrent. In the period of a few months in late 2000 and early 2001 there were two books published on the topic of zero (Kaplan, 2001; Seife, 2000). More recently, an even shorter time-frame saw the publication of three substantial books on the Riemann conjecture (Derbyshire, 2003; du Sautoy, 2003; Sabbagh, 2002). Some other recent texts have focused on a particular branch of mathematics: for instance, Barabasi (2002), Beltrami (1999) and Havil (2003). (I have more than thirty other titles in this category on my shelves alone.)

Other books again look at the discipline more generally – see, for instance, Gowers (2002), Higgins (1998) or Stein (2001). (Again, I could offer more than fifteen others from this same short time period.) A surprisingly large number of authors choose to emphasise the artistic, aesthetic and spiritual connections of their publications in the titles they give their books. See, for instance, *The Artful Universe* (Barrow, 1995), *The Universe and the Teacup: the Mathematics of Truth and Beauty* (Cole, 1998) or *It Must be Beautiful: Great Equations of Modern Science* (Farmelo, 2002) – again, there are over thirty more I could mention, all with this characteristic.

Two further categories with a strongly human flavour are individual biographies or autobiographies of mathematicians such as Flannery (2001), Nasar (1998), Hoffman (1999), Schechter (2000) and Rota (1997), as well as institutional histories such as *Masters of Theory: Cambridge and the Rise of Mathematical Physics* (Warwick, 2003). The strongly human flavour of these biographies is sometimes nothing more than glimpses into the everyday lives of mathematicians, which include – as ours do – families, sickness, vacations and heartaches. These are aspects of a mathematician's life that might have astounded Picker and Berry's schoolchildren.

A relatively new phenomenon is the thematic role played by mathematics in drama, cinema and fiction. Novels with a significant mathematical dimension include Doxiadis (2000), Petsinis (2000), Papadimitriou (2003), Schogt (2000) and Woolfe (1997). Several recent plays and movies have been exceptional for their quality and for their substantial mathematical content. Tom Stoppard and Michael Frayn are among the world's leading playwrights and their respective works, *Arcadia* and *Copenhagen,* have been theatrical standouts. A much younger playwright, David Auburn, won the Pulitzer Prize for Drama in 2001 for his play *Proof,* which grappled with themes of mental illness, mathematical creativity and also generational conflict. Ron Howard's cinematic version of Sylvia Nasar's biography of the Nobel prize-winner John Nash was awarded an Oscar for Best Picture of the Year in 2002. Two earlier productions, *Good Will Hunting* and *Pi,* were also distinguished by the inclusion of significant elements of mathematics and strong and positive audience reactions. [4]

Of course, some of the portrayals of mathematicians and mathematics in the recent flood of literary, dramatic and cinematic productions have been 'classical' in nature. They have, as in the biographies of Erdös or in Russell

Crowe's portrayal of John Nash in the film version of *A Beautiful Mind,* extended and reinforced the stereotype of the mathematician as the 'ultra-geek', a (male) person apart. Of more interest, perhaps, are the female mathematical characters created by David Auburn and Tom Stoppard in their plays *Proof* and *Arcadia.* Part of the message in these latter two cases seems to be that one should not assume that mathematical ability comes only in certain preconceived personifications. Much the same message, with more of a class basis rather than gender, underlies the character of the lead role in the film *Good Will Hunting.*

In a slightly different category is the sensitive depiction of the leading character in Mark Haddon's (2003) highly praised *The Curious Incident of the Dog in the Night-Time.* Christopher John Francis Boone is a teenaged boy who is autistic. Mathematics plays an important part in his life: doubling twos helps him fall asleep, numbers act as human markers much as hair colour and height do for most. Indeed, numbers structure his entire life – as the prime-numbered chapters of his 'autobiography' quickly reveal. In this regard, Haddon seems to be unawarely portraying with considerable success some of the findings emerging from the research of psychologist Simon Baron-Cohen (2003) at Cambridge University (see also Chapter ω).

Long as these lists are, they represent only a partial accounting of a much more extensive collection of materials. I have, for example, only touched the surface of an exceptionally large body of knowledge that examines the intersection of mathematics and art. Indeed, a non-trivial number of artist/mathematicians have written about the mutual inspiration the one discipline has provided in their pursuit of the other. This mutual inspiration is one that the Pythagoreans would have taken for granted.

What sort of explanation can be offered for this undoubtedly burgeoning phenomenon? One that seems feasible is that artists are doing what artists are supposed to do, namely identifying key contemporary issues and exploring new ways of seeing the world. So we have playwrights, movie directors and novelists, as well as sculptors, architects, designers and computer programmers, playing with the central construct of human sensitivity to pattern and form and their abstract extensions, which could serve as a passable specification of a prime motivation for mathematics.

Where Mathematics Might Come from

Perhaps the mathematical perspective most supportive of an embodied genesis for the discipline can be found in the book *Mathematics: Form and Function* by Saunders Mac Lane (1986). In the initial pages of that text, Mac Lane outlined his views on the origins of mathematics. He identifies various fundamental "formal notions", such as cardinal number, continuity, group and topology. He then argued that, "These formal notions arise largely from premathematical concerns which can best be described as 'human cultural activities'" (p. 34).

Mac Lane saw a progression from various types of human activity through the stages of informal "ideas" generated by the activity and then more abstract "formulations". The activity of "collecting", for example, leads to the informal idea of "collection" and is formalized as "set". Similarly, we have "Building, shaping" leading to "Figure symmetry", to "Collection of points" and "Choosing", "Chance" and "Probability" (p. 35).

Mac Lane's emphasis on fundamental human actions and their elaborations, from a mathematical perspective, resonates strongly with an exceptional corpus of work by another scholar in a different field. For some three decades now, Ellen Dissanayake has pursued the study of art in human societies in settings that are geographically and culturally diverse. Then, in three remarkable books she has, almost single-handedly, challenged many of the most longstanding views about human nature. In 1988, at the beginning of *What Is Art for?,* she noted the almost complete lack of serious attention, in most anthropological texts, to the role of art in human cultures. Then, in a series of tightly argued, beautifully illustrated and well-written chapters, she remedied this omission.

In 2000, in her *Art and Intimacy: How the Arts Began,* she traced back the roots of much civilized human behavior to the interactions between parent (in particular, mother) and child. It is, however, in her middle book from 1995, *Homo Aestheticus: Where Art Comes from and Why,* that she revealed perspectives on human predispositions from an aesthetic orientation which overlaps greatly with Mac Lane's list of premathematical human activities.

Dissanayake's perspective is, in many ways, a constructive and optimistic one. She writes with eloquence and conviction about the satisfactions that come from the successful manifestation of the basic aesthetic motivation of "making special". There is, perhaps, merit in acknowledging both the power of Dissanayake's insight and the potential for extensions of her ideas into areas of mathematics.

The role that aesthetics may play in cognition and human action was not a prominent theme in twentieth-century scholarly work, although back in 1968, for example, Gregory Bateson wrote a paper entitled 'The moral and aesthetic structure of human adaptation'. In it, he remarked:

> It is possible that aesthetic perception may be characteristic of human beings, so that action plans which ignore this characteristic of human perception are unlikely to be adopted, and even unlikely to be practicable. (1991, p. 257)

Undeveloped glimpses [5] like this, perhaps unsurprisingly, were also to be found in a number of places in the publications of Alfred North Whitehead.

One of the last public lectures he delivered at Harvard (in his mid-seventies) was entitled 'Mathematics and the good', echoing the title of a famous lecture of Plato's ('On the good') that seems to have dismayed its original listeners by being all about mathematics. In Whitehead's talk, also foreshadowing some of George Steiner's observations fifty years later, we find:

> The notion of the importance of pattern is as old as civilization. Every art is founded on the study of pattern. [...] Mathematics is the most powerful technique for the understanding of pattern, and for the analysis of the relationships of patterns. [...] Having regard to the immensity of its subject-matter mathematics, even modern mathematics, is a science in its babyhood. If civilization continues to advance, in the next two thousand years the overwhelming novelty in human thought will be the dominance of mathematical understanding. (1948, p. 117)

A few years earlier, echoing the opening quotation by Vargas Llosa, he had remarked in his book *Modes of Thought:*

> By reason of the greater concreteness of the aesthetic experience, it is a wider topic than that of the logical experience. Indeed, when the topic of aesthetics has been sufficiently explored, it is doubtful whether there will be anything left over for discussion. (1938, p. 86)

Some fifteen years after Whitehead's death, one philosopher (Norman, 1963) tried to capture an element of Whitehead's philosophical perspective:

> There runs through Whitehead's writing an arresting ambivalence toward one such model [expressing a fundamental decision and orientation with respect to experience], the *mathematical.* To explore his reflections upon mathematical method in philosophy is perhaps to win the most fascinating lesson in the Whiteheadian legacy. A close reading of his remarks upon the subject suggests that he found two uses of mathematics—the one abortive and barren, the other rich and indispensable. We shall call these the *skeptical* and the *aesthetic* use, respectively. (p. 33; *italics in original*)

The contrast between these brief and sporadic comments on the aesthetic and its connections for most of the last century and the concentrated attention it has begun to receive in the last decade is striking.

In addition to the works cited previously, there have been significant publications in physiology (Wilson, F., 1998), in technology (Gelernter, 1998), in philosophy (Levinson, 1998) and in evolutionary psychology (Voland and Grammer, 2003). In Britain, the B.B.C.'s Reith Lecturer for 2003 was the neuro-physiologist C. V. Ramachandran, who gave five lectures on the theme of *The Emerging Brain.*

The third of these was entitled, *The Artful Brain* and in it Ramachandran listed what he called "the ten universal laws of art". Three of his ten 'laws' have immediate and strong connections to mathematics. These are 'symmetry', 'repetition, rhythm and orderliness' and 'balance'. In fact, however, when one reads the elaborations that Ramachandran provides for the remaining seven (peak shift; grouping; contrast; isolation; perception problem solving; abhorrence of coincidence/generic viewpoint; metaphor), it appears that a strong case can be made for the mathematical–aesthetic character of all of his proposed 'laws'.

Homo Mathematico–Aestheticus?

In 1998, the distinguished Dutch-American mathematician and historian, Dirk Struik wrote a foreword for the English edition of Paulus Gerdes's (2003) ethnomathematical treatise, *Awakening of Geometrical Thought in Early Culture*. In this intriguing introduction to a stimulating work, Struik identified three "attitudes", quintessentially human, namely, *homo observans, homo ludens* and *homo laborans*. Consistent with his life-long political convictions, Struik gave pre-eminence in this categorisation to what he termed the "dynamic" approach of *homo laborans* which, as he noted, "is implicit in the Marxian point of view" (p. ix).

At the very end of this section of his foreword, Struik obliquely commented:

> Incidentally, the symmetry and harmony of forms that turn out to be most efficient (many examples appear in this book) also strike us as more agreeable, *beautiful*. A source of the birth of aesthetics? (p. ix; *italics in original*)

In making this brief and speculative comment about aesthetics at the end of his foreword, Struik was behaving in the manner of Bateson and Whitehead. It is, I think, a relatively minor modification of Struik's views, to see all three of his 'attitudes' as close cousins of the (perhaps more fundamental?) *homo mathematico-aestheticus*. We might, for example, ask just which forms and shapes catch the attention of *homo observans,* and why.

Unfortunately, Dirk Struik will not be a participant in any subsequent discussions of these issues. (He was 104 years old when he wrote the preface to Gerdes's book and he died two years later. [6]) His mathematical power and his human wisdom would have made him an especially valuable contributor to a discussion of the numerous and complex, but important and fascinating, connections between the human species and the equally human discipline of mathematics.

Notes

[1] The fundamental nature of the human drive for understanding is perhaps most directly and famously articulated in the opening lines of Aristotle's *Metaphysics,* where he categorically stated, "All humans by nature desire to know". The special place of the subject of mathematics in the history of human efforts to seek out the "inner truth of things" is well documented.

[2] Although originally written about 'the poet' (another outsider group often considered 'odd'), Charles Baudelaire's (1861/1995, p. 16) closing line from his poem 'The Albatross' – *Ses ailes de géant l'empêchent de marcher* ("His giant's wings prevent him from walking") – fits the public image of an awkward, regretfully earthbound (male) mathematician.

[3] The argument for Benjamin Peirce as some sort of archetype does not rest only on the fact that he was a forceful and long-lived individual working in a particularly

influential institution at a crucial period for the formation of attitudes and patterns of behavior. It is a remarkable fact that every President of Harvard University from 1862 to 1933 had studied with Peirce (Menand, 2001, p. 154).

[4] A less well-known phenomenon is that of mathematicians who have written significant works of drama and literature. This number includes John Mighton of the Fields Institute at the University of Toronto, whose play *Possible Worlds* won a Governor General's Award for Drama in 1992, and the University of Maryland's Manil Suri whose *The Death of Vishnu* (2002) is highly reminiscent of the works of prize-winning Canadian novelist Rohinton Mistry (*A Fine Balance, Family Matters*), whose first academic degree was in mathematics. In this category, ring theorist Vladimir Tasić occupies a particularly interesting position, being both a published author of fiction (*Herbarium of Souls*, 1998) but also one of philosophy/cultural studies (*Mathematics and the Roots of Postmodern Thought*, 2001).

[5] Another can be found in Jane Jacobs's book, *Cities and the Wealth of Nations* (1984). She cited the Japanese anthropologist Tadao Umesao's theory that "an esthetics of drift" (as opposed to "resolute purpose" and "determined will") is an effective economic strategy (p. 221). Shortly thereafter she quoted the M.I.T. metallurgist, C. S. Smith's observation that necessity is not the mother of invention. This role, according to Smith, belongs to what he calls "esthetic curiosity" (p. 222).

[6] For more on Struik's mathematical and political life, see Powell and Frankenstein (2001).

Section C

Mathematical Agency

Introduction to Section C

The three chapters of Section C, *Mathematical Agency,* each adopt a more interdisciplinary approach to central aspects of the mathematical enterprise as a human endeavour. Drawing in particular on parallels in the disciplines of history of science and technology, the visual arts, theology and iconographic history, these chapters propose ways in which aspects of the mathematical aesthetic can be understood in terms of broader themes, wider intellectual movements and the nature of human experience.

Nicholas Jackiw, in Chapter 7, deals with the dynamic geometry software environment *The Geometer's Sketchpad,* in which the user creates and manipulates mathematical constructions. When compared with other forms of mathematical representation, *Sketchpad* constructions partake of disembodied geometric abstraction and yet show a responsiveness to our human presence. Many find it compelling to interact with mathematics incarnate. Why this engagement should be found so engrossing is less well understood. In an attempt to locate the historical antecedents and contributory factors to the experience – and pleasure – of creating and manipulating *Sketchpad* constructions, Jackiw positions dynamic geometry within a broader and very old tradition of mechanical devices. This allows him to reveal the unusual combination of aesthetic motivations to dynamic geometry activity which stem, in part, from a *Sketchpad* user's dual capacity to wield magical as well as explanatory mathematical power.

In Chapter 8, David Pimm aligns the term 'aesthetic' with its original Greek meaning of "pertaining to the senses" and that of 'theorem' with "that which is seen". He contrasts the mathematical use of text and image (the written with the drawn) and examines the role images have played in mathematical argument. This discussion is enhanced by drawing on the writing of early twentieth-century European visual artists (such as Kandinsky and Malevich) about 'abstraction'. Basing part of his chapter in an account of the Bourbaki group's attitudes to the mathematical image, Pimm takes seriously member Pierre Cartier's observation that 'The Bourbaki were Puritans' and traces links of both attitude and argument to the Reformation sixteenth- and seventeenth-century banishment and even destruction of icons.

Dick Tahta, in Chapter 9, offers a sequence of broadly connected thoughts about 'sensible objects'. Here, the adjective 'sensible' may suggest what can be perceived by the senses or what can be understood in the mind. Objects may acquire special significance for specific groups. But this making special inevitably involves controversy: examples include disagreements about the purpose of neolithic stone balls (and later art objects), eighth-century Byzantine iconoclasm and different twentieth-century views about psychoanalysis. Specifically, mathematics and mathematicians exhibit some particular attitudes to the 'objects' they consider. These may become iconic in the original Byzantine sense: there are some fruitful psychoanalytic accounts of how this might happen. But, in the end, the nature of the mathematician's objects – like those of many others – remains a mystery.

CHAPTER 7
Mechanism and Magic in the Psychology of Dynamic Geometry

R. Nicholas Jackiw

> The dilemma posed all scientific explanation is this: magic or geometry? (Thom, 1975, p. 5)

Draw an arbitrary quadrilateral—or better, build a picture of one in your imagination—and connect its sides' mid-points to form a second, inscribed quadrilateral. If your hand is steady, or your mind's eye clear, this inner shape appears not only quadrilateral, but its opposing sides may appear parallel as well. Are they really? Now, reach with imaginary fingers into the figure itself and grasp one of the quadrilateral's outer vertices. Push it slowly inwards, flattening the original shape like a cardboard box collapsing. The two edges adjacent to your vertex grow straighter and straighter, until eventually they form a smooth, unbroken line and your original quadrilateral has degenerated into a triangle. (Your dragged vertex and its neighbors are collinear.) Across this transition, what became of that inner shape? In its present, extreme configuration, are any secrets revealed? Can you find evidence for why it must also form a parallelogram in your original—and in every possible other—starting configuration?

Now switch to the complex plane and consider there both a circle T and its image T' under some cubic mapping $f(z) = (z - a)(z - b)(z - c)$. If T encircles a, then T' must loop around the origin, since the origin is the image of each root under f. Now imagine expanding circle T so that it grows toward b, while continuing to encircle a. By the time T consumes b, T' will have had to wind twice around the origin (or even three times, if T encircles a, b and c simultaneously). Moreover, since f is analytic, T' will have to evolve continuously from one loop, to two, to three as your circle T encompasses more and more roots. In other words, even before you capture a new root on the plane with T, T' will have spawned a twist in its contour that can self-intersect and wind about the origin an additional time as soon as you arrive at the root. But, in this scenario, how has T' possibly anticipated your intentions for T's future deformation?

The Dynamic Geometry Experience

Users of software such as *The Geometer's Sketchpad* will already be familiar with the type of conjectured activity described here and regard its natural habitat as the computer screen rather than the screen of one's imagination. The ingredients of a typical interaction with such software are those of the above thought experiments: a motivating mathematical curiosity about shape or spatial relationship; a quick diagrammatic construction of that configuration's basic skeleton; a more probing exploration of that shape's kinematic behavior when manipulated dynamically.

The mechanical heart of such a 'dynamic geometry' experience lies in its idealized and friction-free mathematical physics, that permits you to vary a constructed figure along every possible degree of freedom left by the rigid mathematical definitions that you yourself have asserted as essential to your figure. In dynamic geometry, mathematical figures are completely unlike conventional static images that act as simple illustrations or representative diagrams. They are new mathematical objects that map a temporal past of specification and definition onto a present graphical configuration and also onto a future potential for manipulation and possible constrained response. Shape becomes infinitely plastic within the software: your *arbitrary* quadrilateral permutes into *any possible* quadrilateral, into *all possible* quadrilaterals. In the close mathematical feedback loop of haptic cause and visual effect, between dragging a vertex and the diagram transforming, attention naturally shifts from specific shapes and configurations to the relationships, harmonies and proportions that hold among shapes and across their endless configurations. In dynamic geometry, these invariant properties appear as the landmarks and sign-posts that orient the map connecting the specific to the general, the concrete to the abstract.

This is a powerful mathematical experience and a vivid one. In the field of geometry, the compass and straight-edge have long demonstrated the close link between physical tools and epistemological ones (see, for example, Scher, 2002, or Bartolini Bussi and Boni, 2003). Within that tradition, dynamic geometry tools seem startlingly new to the many different communities that involve themselves in geometric thinking and practice. Though the idea—dynamic to its core—of infinitely malleable shape is far more easily communicated and understood through demonstration and experience of the software than through reading words in static print, a broad literature has grown up around the software over the past dozen years. Surveying this writing, Schattschneider and King (1997) conclude:

> [Not only has] dynamic geometry software [...] had a profound
> effect on classroom teaching wherever it has been introduced [but
> it] has also become an indispensable research tool for mathematicians and scientists. (p. ix)

The breadth of this accord—which is echoed across and throughout the literature—is rather curious, situated as it is straddling a diversity of communities. Both professional mathematicians and middle-school mathematics students have previously relied on software tools in pursuing their respective vocations, but rarely, if ever, on the same ones.

If the appeal of dynamic geometry routinely spans such divides, is it by tickling some superficial sweet tooth or by plucking some deeper experiential chord? I am intrigued by how often, in first-person accounts of their encounters with *Sketchpad,* writers invoke gripping suspense and delighted surprise as essential characteristics of their experience. While a single community might foster a shared, critical perspective on the software's potential for impact, the common voice sounding from *Sketchpad's* broader user communities is less intellectual than emotional. It suggests an almost organic allure to dynamic geometry activity—to interacting with mathematics at a level that is simultaneously palpably virtual and virtually palpable. Writing as a mathematician self-critically enumerating various factors and conditions contributing to specific engagements with the process of mathematical discovery, Douglas Hofstadter (1997) invokes this more psychological dimension of the software's role:

> One further key factor that mustn't be overlooked is the fortuitous existence and tremendous power of *Geometer's Sketchpad.* Somehow, this program precisely filled an inner need, a craving, that I had, to be able to see my beloved special points doing their intricate, complex dances inside and outside the triangle as it changed. (p. 13)

This aesthetic satisfaction is not simply an effect of novelty and first impressions—the sudden pleasure of encountering an unanticipated, bold idea. Dynamic geometry 'experts' of long familiarity with *Sketchpad* describe similar deep, affective responses. Reflecting upon his years not only pursuing mathematics with the software, but also researching educational issues and designing curriculum surrounding its use, Daniel Scher (2003) writes:

> *The Geometer's Sketchpad* software […] has cast aside any nostalgia I might feel for my long-departed compass. I suspect [this] is a familiar story to many readers. On an instinctive level, building interactive geometric figures feels […] sound. It is also plain fun. (p. 36)

As the designer of *The Geometer's Sketchpad,* and as co-coiner of the term 'dynamic geometry' to describe it, I am fascinated by dynamic geometry, by the nature of its appeal, by the diversity of actors for whom it is tangible and by the range of contexts in which it finds voice. As a student of the history of software-specific ideas, I am intrigued by the way in which variations of *Sketchpad's* core dynamic geometry idea arose semi-independently in the space of a few years and quickly colonized the globe. [1]

This curiosity distinguishes between the idea of dynamic geometry and the code in which it finds form. While as a paradigm for exploring mathematics, dynamic geometry seems vital and durable, actual computer programs pass brief and brittle lives before hurtling toward technological obsolescence. Yet somehow these ephemeral software shells enclose the vivid, robust heart of the dynamic geometry idea. Since the pretense of we programmers who claim to be software 'designers' is that we aspire, if not to art, then at least to the pursuit of some discipline of formal values that inform and illuminate our more mechanical undertakings, these perspectives sharpen my interest in dynamic geometry's appeal.

Of course, not everyone's encounter with dynamic geometry leads to the enthusiasm of the reports I quote here. And so, if I can better understand that appeal where it is present, then as a programmer I can perhaps better bottle it for others; I can commodify it. But as an author or inventor working in the medium of code, I struggle with these questions at more subjective, inchoate levels. As I make additions or changes to a program—a feature here, a sub-routine there—I must ask if they amplify its dynamic geometry nature or if they suppress it. If the program is an instrument, and dynamic geometry its music, in what direction lies best tune?

First and foremost, these are aesthetic questions and pursuit of their plausible answers takes us quickly into murky psychological terrain and conjectural evolutionary causalities. Understandably, neither the mathematical nor the educational literature embraces questions that are so baldly subjective and psychological. But in attempting at least to acknowledge the apparent broad impulse toward, and appeal of, these software tools, certain recurrent ideas in that literature nonetheless suggest that dynamic geometry is somehow natural or obvious or inevitable.

One sees such a treatment, for example, in the recurring proposition that *the great mathematicians of history*—Newton is summoned here, or Apollonius—*have* always *"thought dynamically"*. The idea proposes that, since dynamic geometry software use encourages the development of some visual form of mathematical intuition, it helps practitioners see (on their computer screens) in the manner great mathematicians see (in their imaginations). Conceits such as these are provocative and useful—they have allowed dynamic geometry technology to pass unchallenged by many gatekeepers of acceptable mathematical or didactic practice—and they are perhaps even correct. Hadamard (1945), for instance, wrote of the deep importance of visual and kinetic thinking in the creative processes of eminent mathematicians, and even of their dominance over other modes of speculation.

But while one readily believes that mathematicians care about their ability to 'see' like great mathematicians, and that at least some teachers hold comparable beliefs, it is far more difficult to presume that younger students come pre-equipped with similar passions. Blunt evidence on student attitudes towards mathematics and mathematicians [2] suggests that many of today's students—at the secondary level at least—might rather pluck out their own eyes than see like a mathematician. Thus, these propositions, while in some sense explaining or justifying the role of the software or its presence in a particular context, seem to me poor placeholders for more explicit hypotheses about dynamic geometry's aesthetic appeal or the motivation or impulse toward it.

A second line of proffered explanation for dynamic geometry's seductive naturalness often focuses on its position within the technological tradition of computers in geometry. From the mathematical perspective, strong affinities between geometry's particular combination of deductive rigor and visual appeal and the computer's capacity for dead certainties, hard labor, and—more recently—striking imagery have given rise to a long history of geometric modeling, investigation and visualization realized in whole or in part with digital machinery. Since at least the 1950s, software researchers have pursued various automated models for geometric theorem proving. Paralleling the emergence of these technologies enabling experimental mathematics has come a vision of the potential for computer-based geometric modeling to transform human mathematical activity.

In this reading, dynamic geometry is but one of an evolutionary line of computational geometry technologies—say, the symbolic geometry technologies of the 1960s, the first graphical images of the 1970s, the fractal geometries of the 1980s and then the dynamic geometries of the 1990s. But while this rough overall chronology is indisputable, the fact of its presence in a larger frame does not vacate the interest or relevance of its substantiating elements. In other words, the relation of today's geometry software to other geometry software sheds no immediate light on the particular pleasures of dynamic geometry or on the intensely visual and physical processes of building, manipulating and *wielding* continuous permutable geometry on-screen and in real time. These pleasures form the essential particulars by which dynamic geometry differs from its immediate software antecedents, rather than the common soil from which they all sprout.

Didactic Mechanism

I believe at least part of this disconnect—between ample descriptions and absent explanations (or even explorations) of dynamic geometry's aesthetic import—reflects less the tendency to inscribe dynamic geometry too quickly within existing technology traditions than the tendency to place it only within technological traditions of insufficient scale and depth. The broad social commotion of recent decades about new and novel digital technologies of all shapes and colors has so relentlessly focused on the future—on utopic and dystopic visions of information-age, computer-based, high-tech, multimedia, virtual, on-line, dot-com e-futures—that it has obscured at times our ability to look deeply into the technological past. Thus, the dynamic geometry discourse has largely (though not completely) overlooked the rich human history of older machineries and pre-computational technologies that—like dynamic geometry—have been deployed as 'working models' of mathematical and physical properties and as conceptual illustrations or exemplars across the natural sciences.

Historically, innovation in technological devices for didactic purposes has accompanied (or, often, preceded) corresponding technological advances in applied or utilitarian domains: for each new calculating or problem-solving technology (for each new abacus, say, or astrolabe), there has been the corresponding new technological demonstration or pedagogical manipulative (a locus linkage here, an armillary sphere there). When the idea of educational technology is interpreted against a history not of recent decades but of millennia, these didactic mechanisms are its proper object. It is within this broader view, I argue, that we can find useful theoretical perspectives on, and greater resonance with, the dynamic geometry experience.

Of course, to read dynamic geometry within a history of mechanical devices, and of the traditions and philosophies that have inspired them, is to take giant footsteps away from present context (from the local minutiae of computer features; from the specific social, educational and psychological contexts in which dynamic geometry has been coined and used) as to risk abstracting or generalizing away all necessary particulars. But there are strong reasons to continue. If the tradition of didactic or demonstrative mechanism puts forward the machine both as a model of patterns of nature and science and as a means to reveal and understand those patterns, then both geometry and dynamic geometry fall squarely within that tradition. Few domains offer as close an equivalence between the foundational ideas or principles of the domain and the mechanical devices that embody them as does Euclidean geometry. Textbooks call it 'compass-and-straight-edge' geometry to note that, here, the constituting ideas *are* the various defining forms of mechanical apparatus. The dynamic geometry experience, which begins with users familiarizing themselves with electronic simulations of the originary mechanical compass and physical straight-edge, reinforces this mechanical orientation and even compounds it. To the extent it represents a *double* mechanical mediation of geometry, with modern software tools representing ancient drafting devices in turn representing abstract and transtemporal mathematical ideas, dynamic geometry can be said not only to uphold, but almost gleefully to celebrate, the tradition of mechanism.

Trivializations of the mechanical

But what might such a tradition of didactic mechanism tell us about dynamic geometry or, for that matter, about other ways of knowing? From the present perspective, deep within the digital era, the answer might first appear to be 'not much'. When considering the history of mechanism—and especially that of explicitly demonstrative, didactic mechanisms—it is tempting to mistake an earlier era's predisposition to the mechanical, and to mechanical forms of explanation, as implying an era of only simpler or more obvious truths. The promise of exposed clockwork denies the possibility of hidden meanings and unseen movers. As scientific epistemology, it is distinctly pre-psychological, pre-quantum-mechanical, in its imputation that fundamental structure can always be brought to the surface: to expose a phenomenon is to explain it.

Current scholarship often adopts this sort of confining interpretation when it locates and limits the tradition of demonstrative mechanism to within the temporal and ideological bounds of the discourse of mechanistic philosophy. This is, of course, the intellectual movement that fueled the scientific revolution and much of the instrument- and machine-dominated imagination of the Enlightenment. In this philosophy, scientific understanding is built upon the idea of the self-regulating and self-governing clockwork, rather than around the more opaque teleologies of vitalism and scholasticism. From its seventeenth-century roots, the movement flowered quickly. By the middle of the eighteenth century, for example, in his tract *Man a Machine,* the atheist physician Julien de La Mettrie (1748/1912) could read all of human physiology and physiological homeostasis through directly mechanical conceits:

> The human body is a watch, a large watch [...] If the wheel which marks the seconds happens to stop, the minute wheel turns and keeps on going its round, and in the same way the quarter-hour wheel, and all the others go on running when the first wheels have stopped because rusty or, for any reason, out of order. (p. 141)

This philosophical backdrop, in which imagined or metaphorical machines serve as intellectual emblems of self-evidence and self-governance, and in which machines serve as a basis or model of insight and explanation, is often seen as emerging in response to growing technological sophistication of utilitarian machinery—of time pieces, navigational instruments and the engineering apparatus of agriculture, architecture and warfare. In other words (the reasoning goes), machines must achieve a certain complexity of function in order for a description of their workings to achieve the status of intellectual explanation. And once that complexity has been broadly achieved or perceived, it becomes natural to imagine actual machines being built—by philosophers, scientists and educators—as explanatory devices demonstrating simpler or more accessible scientific principles. Thus, proceeding down this line of thought, one can limit the role of the actual demonstrative mechanism to particular didactic contexts within the era of mechanistic philosophy's currency, commencing perhaps with the scientific revolution and petering out of steam, as it were, with Watt's engines two centuries later.

Popular or non-academic conceptions of the tradition of mechanism are hardly more generous in the role and influence they accord to didactic mechanical devices. Present-day fiction and film frequently limit their portraits of mechanism—particularly classical, pre-electronic mechanism—to the distinctly improbable and the decidedly unreliable. With the occasional exception of the various technologies of violence and war, the dramatic purpose of the machine is either to function against all possible odds (the Rube Goldberg mechanism) or to malfunction vividly, just at the moment it becomes essential ("Captain! It's jammed!").

Caught between these academic and popular conceits, our present-day images of yesteryear's mechanisms are to be forgiven some degree of pejorative stereotyping. The orrery of one's mind's eye may be ornate in its brass gearings, but alas no longer functions. Though it is lovely—mounted on its mahogany pilaster—it seems also slightly precious and perhaps a touch absurd. We delight in finding such machines well-preserved in science museums, but otherwise do not miss them. Along with linkages, screws and inclined planes, they evoke a time when known physical laws still bore a correlation to reasonable intuition, when theories sought to simplify rather than complicate phenomena and when a working facility with the principles of engineering fell well within the realm of the aspiring polymath. Confronting these simple bygone mechanisms from our contemporary, post-modern vantage—from this era of complex systems and non-linear dynamics, from so far along the information highway—we interpret their intent to simplify as simplistic and their desire to move structure toward the surface as only—or inevitably—superficial.

This would be a grievous mistake, for these conceptions—scholarly and popular—are almost entirely wrong. To relegate the impulse toward demonstrative mechanism to an Enlightenment sense of didacticism incorrectly imagines the timetable of such technologies' evolution (and demise) while also underestimating their intellectual import or moment across that timetable. First, I argue, the attention such misconceptions pay to the scientific revolution and the Enlightenment overlooks a much broader cultural and historical matrix of demonstrative mechanism. Second, in heeding only the didactic and ostensibly educative role of demonstrative mechanism, such misconceptions ignore the signal fact that—across this broader history and in all moments within it—demonstrative mechanisms were deployed as often to confound as to reveal. They bear witness to sorcery as often as to science and they stand in esoteric, as often as populist, relations to human truth and knowledge. Any account of dynamic geometry in the tradition of didactic mechanism must equally account for magical mechanism, for they are one and the same.

Meaning and Mechanism

In this section, I examine one by one my objections to a 'weak' interpretation of didactic mechanism. I have suggested that modern popular representations tend to portray classical mechanism as inevitably a preposterous contraption: that is, as parody. And parody obviously has dramatic traction only against some prior, contrary impulse, some unspoken presumption of relevance, perhaps, or gravitas. But the same requirement of prior context must apply to mechanistic philosophy's use of the idea of machine. The view I criticize holds that such a philosophy only becomes possible after an imagined point of sufficient technological advancement. In other words, La Mettrie can conceive of man as machine only once machine has become as rich a locus of complexity and potential as man. But how do such attributes come to endow our conceptions of 'prior' machinery? Where is one to find them?

To say such philosophy becomes possible only after mechanical technologies have reached a high level of sophistication and potential is to demand context—sophistication and potential—that are not properties of the machine as object, but rather are human conceptions embodied in discourse and cultural practice around the machine as idea. Machines co-emerge with the meanings we impute to them: there has been no era of unmediated technological production and our machines can only be said to 'possess potential' once we are committed to the possibility of such an idea. Thus, the particularly heightened form of mechanistic philosophy found in the Enlightenment must be inscribed in a larger or looser mechanistic intellectual tradition, one that parallels, fuelling and being fueled by, the entire history of machine-making, the whole course of human technological development.

No one documents the scope of this deeper and more epic mechanistic tradition—of machines being conceived and received as embodiments, exemplars, repositories and demonstrations of profound scientific knowledge—more clearly than the historian of science Derek de Solla Price. Price (1964) traced the "strong innate urge toward mechanistic explanation" (p. 10) back through antiquity to the very edge of pre-history. In the talking statues and articulated dolls of Middle Kingdom Egypt, in jointed African transformation masks and traditional Indonesian puppetry, he located the earliest progenitors of the classical navigational device, the timepiece of the middle ages, the Renaissance instrument, even the industrial machine and modern computer. Across this history of machinery, explanation, novelty and demonstration—more than, say, utility, industry or profit—act as the compelling motors of technology development. As Price observed:

> the most ingenious mechanical devices of antiquity were not useful machines but trivial toys. Only slowly do the machines of everyday life take up the scientific advances and basic principles used long before in [...] overly-ingenious, impracticable scientific models and instruments. (p. 15)

Dynamic Geometry as mechanism

How does dynamic geometry fit into the far vaster landscape of demonstrative mechanism? Of course, in one sense, the stretching line segments and spinning circles of the typical dynamic geometry configuration already resemble the rods and pulleys, the pistons and axels, of classical mechanical devices. Thus, a common pastime of the dynamic geometry scholar is to recreate (within the software environment) the many marvelous linkages and drawing devices by which the ancients produced their conic sections and their many other curves. (See, for example, Scher, 2002, or Dennis and Confrey, 1997.) Perhaps this pastime responds to the way in which the most common form of historical record of these devices—namely, their schematic diagrams and their mechanical blueprints—are closer to mathematical figures than to actual machines and so almost cry out for kinematics, for their missing actual dynamics. Our conditioned responses to a particular graphical configuration or diagram may be entirely different if we perceive it as depicting 'an angle from the horizontal' rather than, say, 'an inclined plane'. The former conception is static, while the latter entails an obvious, latent dynamics: we push things up inclined planes, we roll things down them. Dynamic geometry is an expressive language within which to frame and phrase such conceptions. In this sense, dynamic geometry interpretations of mechanical schemata and motion diagrams make such representations appear more true to themselves, as well as to the motions and movements they attempt to describe.

But this correspondence of dynamic geometry to classical mechanism takes mechanism only literally and not yet metaphorically. The distinction is important, for the metaphoric meanings we have ascribed to mechanisms over time are far grander than the machines we have built, with the *idea* of mechanism often far more gripping than the machines themselves. For example, in Diderot's *Encyclopédie,* D'Alembert's entry for 'equation' spans many pages. The definition of an algebraic equation as one might understand it today is quickly dispatched and the remaining exposition is devoted to the description of an astonishing mechanical device that can graph all possible rational functions of any conceivable degree. The given specifications are extensive and comprehensive and include some illustrations of the machine (*Le Constructeur Universel d'Equations*) in action. [3] Only at the end of this impressive description do we learn that, due to the tragic constraints of friction, the machine is impossible to build for curves of degree higher than two!

In a mechanically similar vein stands the long search for, and often-claimed invention of, machinery capable of perpetual, self-generating motion. Ord-Hume (1977) surveys the wealth of late-nineteenth-century blueprints and specifications and even scale—though, of course, not working!—models that so inundated the US Patent Office's application process that legislative acts were eventually required to discourage future applicants. In these examples, though the mechanism itself does not function, the idea of mechanism functions clearly—as a placeholder for convincing demonstration, and as psychological guarantor of physical plausibility against the suggestion of incredible or uncredentialable conjecture.

Duality in Mechanism and Dynamic Geometry

This idea of mechanism, this psychological dimension and presence to it, underlies much of Price's account of the two essential domains and applications of early demonstrative mechanisms: the biological and the astronomical. From the dawn of our first technologies, it seems, we have built models of man and models of the cosmos. Thus, on the one hand, we have articulated dolls, statuary, automata and robots; on the other, spinning globes, anaphoric clocks, orreries and planetaria. Price's work has traced the linked technological trajectories of both domains and demonstrated how through even to modern times, the techniques and engineering principles of fine mechanism have appeared first in models within these two domains before migrating to other machined applications. In summary, he observed:

> [Biological and astronomical devices] go hand-in-hand and are indissolubly wedded in all their subsequent developments. In many ways they appear mechanically and historically dependent one upon the other; they represent complementary facets of man's urge to exhibit the depth of his understanding and his sophisticated skills by playing the role of a do-it-yourself creator of the universe, embodying its two most noble aspects, the cosmic and the animate. (1964, p. 15)

I find Price's proposal intriguing as a framework in which to chart some of the less obvious dimensions of dynamic geometry's felt experience and perceived meaning. Certainly, the psychological juxtapositions of Price's dualism are attractive. The biological stands in relation to the astronomical as the two sides of the pan-historic coin of metaphysics: self to universe; I to other; here-and-now to everywhere-always. Evolve this dualism to the present era, however, and we discover that the astronomical universe, though perhaps expanding physically, has contracted dramatically in its psychological import. We can now see with powerful devices to its very limits. In under three hundred years, all of space has collapsed from Newton's unknowable *chaos infinitum ex atomis* to Einstein's handful of elegant equations. Given this, can one perhaps imagine, in Price's dichotomy, the role of the astronomical being displaced by, or extended by, the purely mathematical, as the new nexus of infinity and unfettered possibility?

If such substitutions—of infinity for eternity, of mathematics for *mathesis* —are legitimate, even if only as contemporary shadings or late-modern symptoms of Price's more timeless dichotomy, then the dynamic geometry experience must partake of this same metaphysical tension. This is because it is precisely the milieu in which the individual 'touches' raw mathematical ideas, where personal volition and physical exertion can have seismic impact on disembodied abstractions. Davis and Hersh (1980, p. 36) describe the writerly practice of their 'ideal mathematician' as follows: "His writing follows an unbreakable convention: to conceal any sign that the author or the intended reader is a human being". Dynamic geometry experience is the direct opposite of this, with constant reinscription of the human into the idealized mathematics. It is the dragging of the user's hand—a co-ordinated feat of bone and muscle—that claims the center stage and from which all meaning flowers. Rather than erase the mathematical ego, dynamic geometry inflates it. As the mediating technology by which Price's biological hand touches the astronomical or mathematical *noumenon*, dynamic geometry provides the metaphoric lever by which Archimedes might move the world.

But this trajectory pursues Price at his most conjectural and metaphysical. Even at a less ambitious level, Price's formulation is provocative in its attribution of essential meanings to the *invention* of mechanisms rather than to their *application* or (practical or philosophical) purpose. Thus, to refocus only slightly, the purpose of demonstrative mechanism may be *to have been demonstrated* as much as *to demonstrate.*

We need neither entirely accept this reformulation, nor concern ourselves with clarifying intended purposes from received or extracted ones, to be able to recognize the useful attention it draws to two very different functions of the actual demonstration of the demonstrative mechanism: that is, two functions of the machine in use. The first function—which I have used until now as definitional—is to establish or communicate the essential physical properties and workings of the mechanism and to reveal the scientific principle, the natural law, the inner logic or the outer reality embodied by these properties. But the second function of the mechanism's demonstration is to attest to the ingenuity, skill, authority and potency of the mechanism's creator or wielder. Where the subject position of the first function is indirect and passive—one is demonstrated to—the subject position of the second is immediate and active: one demonstrates.

These two human roles—demonstrator and demonstratee—form the opposing faces of dynamic geometry activity. As one wields the dynamic compass, declaratively summoning circles into existence, one *demonstrates* with a voice almost infallibly potent, almost Euclidean or Cartesian in its certain authority. But the switch from declaring geometry to dragging it about on one's screen recasts one completely as the student, the demonstratee. Here, one's actions are inquisitive and usually tentative: one is seeking, rather than stating. As key components are dragged, the responding motions of one's figure always appear as improvisational choreography—a single possible performance drawn from the limitless configuration space of the mathematics spread across the stage.

These two experiential notes form the particular intellectual and physical rhythm of dynamic geometry work. There is a constant alternation between declarative acts of construction and interrogative acts of dragging. There is a tidal movement between the modality of the certain, the expository and the declamatory on the one hand and that of the tentative and conjectural on the other, occasionally even the suddenly startled or the frankly surprised. These are the motions behind the aesthetic pleasures of and affective rhythms to dynamic geometry activity described in the literature—behind Hofstadter's "intricate, complex dances" and Scher's sense of "plain fun". Demonstrator, demonstratee; actor, audience—one might switch dozens or hundreds of times between these two modes in a single dynamic geometry episode. They comprise its magnetic poles.

Magic and explanation

The resemblance of these two roles to the classic didactic relationship—demonstrator to demonstratee as teacher to student—is hardly accidental. Positioned at the intersection of these roles, and acting as the *communiqué* between them, the demonstrated mechanical device itself is the site of a transaction of unequal authority and power. When the mechanism is deployed in a didactic or explanatory mode, the function of the demonstration is to harmonize this disequilibrium. By contrast, when the device is deployed in a more magical mode—when its purpose is to astound and amaze, rather than to explain or enlighten—its function is to sharpen that divide, to emphasize the inequity. Actual mechanical demonstrations (like actual uses of technology) alternate between the two modes, between harmony and disequilibrium, between explanation and magic.

Though mathematical practitioners who embrace dynamic geometry may cringe at, or reject outright, a sense of magic at the core of their technological practice—for, in a simple view, their mission is that of banishing magic through rational explanation—nonetheless, I think it is the most just word. Revelation and obfuscation partake of the same psychological substance: the idea of the sage moves only slowly from the sorcerer to the scholar. [4] After all, magic, even more than pedantry, relies on the presumption and, indeed, demonstration of transparency.

Magic long haunts our experience of geometry. Though we may no longer pursue the mystic practice of the Pythagoreans in our classrooms, perhaps we retain a taste for them. The best-known proof of the theorem of Pythagoras, attributed to Bhāskara, still comes labeled only with the sorcerer's imperative: *Behold*! And magic long endows our experience of technology. Today's robot-making engineers descend directly from the architects of ancient Egyptian temples, where Hero (62AD) tells us wine poured from ewers untouched by human hands, doors swung mysteriously open and shut and stone idols spoke in incomprehensible tongues, all in response to unseen, barometric mechanisms actuated by priests' careful regulation of the sacred fires.

Fulfilling both of these traditions, dynamic geometry is itself rich in magic. There is magic in its simplified and idealized physics. In dynamic geometry animations, perpetual motion is a pedestrian state of affairs, not some mad alchemical dream. D'Alembert's machine functions properly in this software world, unlike in our own, for curves of any possible degree, and Peaucellier's linkage draws a genuinely straight line for the first time. There is magic in the causality of dynamic geometry. A nearby circle's radius adjusts immediately to reflect changes to some far-off segment—despite an utter lack of Aristotelian connection between them. A geometric point clings with parasitic tenacity to its shifting, restless host through no natural law or physical obligation, but rather because a mathematical voice has intoned "let it be so".

And finally there is magic in the very gestures and incantations by which one directs dynamic geometry activity. Perhaps the two most commonly issued commands in *Sketchpad* are the command to Hide, and its opposite, the apotheosis of *deiknume* proof, to Show All Hidden. These commands are used by teachers to structure and present material coherently, by students to simplify and direct their focus, by mathematicians to explore and to manage complexity. Lastly, the command to set mathematical objects into dynamic geometry is Animate—Dr. Frankenstein's own imperative voiced from the laboratory cauldrons of science, technology and magic. If we overlook what powerful spells these words cast, it is only because we are already comfortable in our role as potent adepts.

Thus, we arrive again at our previous conception, of an experience almost paradigmatically split. The didactic or explanatory aspect of dynamic geometry concentrates and accentuates a depersonalized conception of mathematical knowledge—a mathematics that the language of didactic mechanism presents as self-evident and autonomously functional. But simultaneously—and from a psychological perspective, almost schizophrenically—the magical aspect of dynamic geometry reintroduces and constantly re-emphasizes the acting human self who stands in relation to that disembodied knowledge, as witness, as collaborator and, most powerfully, as omnipotent creator.

These split psychological conceptions of the dynamic geometry subject position mirror opposing aesthetic motivations to dynamic geometry activity. Operations in the explanatory mode invoke the critic's act of aesthetic detachment, of the willful transformation of subject to object. But dynamic geometry's more magical manœuvers invoke the etymologist's understanding of aesthetic engagement, as a full sensory immersion in experience. These dimensions to its felt experience elevate dynamic geometry activity beyond the simple, utilitarian process of drafting animated diagrams of mechanical linkages. They transform it into a practice in which one tangibly manipulates not only mathematics, but also one's own image of the self's relation to mathematical curiosity, understanding and production. Our dragging hand invests the causality and consequence of mathematics with our own almost conspicuous agency. Reciprocally, the ability to influence and impact austere and heightened mathematical knowledge so directly, so immediately, in turn elevates the status of that agency—elevates our own status, as sage and as mathematics' master.

One subconscious voice at the heart of the dynamic geometry experience calls us out of ourselves, into detached, aesthetic contemplation of a plane of pure mathematics. An equal, second voice sings the song of self, celebrating our majesty, our mastery over that dominion. The deep impulse to dynamic geometry, and the rhythm of its activity, arise from these voices' twin duet, from the projection of our volition into geometry's atemporal abstraction. Dragging mathematical objects exerts our influence in that hyperion, building and constructing them asserts our ego—our ability not only to touch that reality, but to shape it, to define it. In this environment, the computer mouse in our hand and the dynamic geometry cursor on our screen are the prostheses of our will and imagination. They form the concentrated and distilled simulation of our self that we push before us, beyond our shadow world and into geometry's clear, platonic light.

Notes

[1] *Sketchpad's* origin, in a National Science Foundation project in the USA, dates from the late 1980s (Jackiw, 1991, 2000). At the same time, and independently, a group of French researchers invented *Cabri-Géomètre,* a similar program involving direct manipulation of geometric figures (Laborde *et al.,* 1989). By 2003, there were at least fifty programs in the same vein, available on the internet, coming from groups across North and South America, Europe and Asia.

[2] See, for example, Picker and Berry (2000), where an internationally broad spectrum of students asked to draw pictures of their ideas of (non-specific) mathematicians create a harrowing portrait gallery of sadists, misanthropes and monsters.

[3] I am indebted to Jean-Marie Laborde, and through him to Michel Carral and Roger Cuppens, for bringing D'Alembert's marvelous machine to my attention.

[4] Within Anglo-Saxon traditions of mathematics and magic, Frances Yates (1969) emblematizes this historical transition in the Elizabethan figure of John Dee (1527–1608). Scholar, bibliophile and founder of Rosicrucianism, Dee stands between the hermetic, magical knowledge of H. Cornelius Agrippa and the alchemists, and the didactic, transparent knowledge of Robert Fludd and the mechanical engineers. According to Yates, it is as much Dee's interest in Euclid as in more occult philosophers that led to popular accusations against him of 'thaumaturgike' (the making of wondrous mechanical marvels, one of the branches of mathematics Dee listed in his 1570 'Mathematicall Præface' to the first English translation of Euclid) and to the pillaging and destruction of his magnificent library.

CHAPTER 8
Drawing on the Image in Mathematics and Art

David Pimm

Studies on the foundations of mathematics and mathematical
method should make substantial room for psychology, indeed even
for the aesthetic. (Henri Lebesgue, 1941, p. 122)

Woe betide him who relies solely on mathematics.
(Wassily Kandinsky, 1931, p. 351)

Aesthetic considerations concern *what* to attend to (the problems, elements,
objects), *how* to attend to them (the means, principles, techniques, methods)
and *why* they are worth attending to (in pursuit of the beautiful, the good,
the right, the useful, the ideal, the perfect or, simply, the true). I have deliberately framed this specification in general terms, so that it applies equally
well to mathematics as to art, historically the realm of much discussion of
things aesthetic. The remarks that comprise this chapter are, to some degree,
organised around these three categories of the what, the how and the why.

Almost as if prophesying the Bourbaki project which commenced just a
few years later (and which forms one of the focal points of this chapter),
Wolfgang Krull (1930/1987) noted [1]:

Mathematicians are not concerned merely with finding and proving theorems, they also want to arrange and assemble the theorems so that they appear not only correct but evident and compelling. Such a goal, I feel, is aesthetic rather than epistemological.
(p. 49)

Krull's sensibility to a distinction between 'the correct' and 'the compelling'
allows a claim that the *aesthetic* goal of mathematical organisation can be seen
to have permeated deductive mathematics since its early Greek inception. [2]

When William Thurston (1994, p. 162) claimed that the job of the mathematician is "finding ways for *people* to understand and think about mathematics", he was going one step further than the common mathematical
view embodied in Samuel Johnson's peremptory retort, "Sir, I have found
you an argument; but I am not obliged to find you an understanding".
Following a discussion about some challenges of machine computational
proofs in mathematics, Yuri Manin (1977, p. 51) offered the maxim that "a
good proof is one which makes us wiser". In doing so, he was not only
making an aesthetic claim about the nature of 'the good' in mathematics, he

was also tacitly echoing a centuries-old distinction between wisdom and understanding (for discussion, see, for example, Read, 1960, p. 17).

Claude Chevalley, when looking back over his many years of involvement with the Bourbaki group since its inception, echoed this aesthetic:

> One mustn't forget that it was Bourbaki who introduced axiomatization into France. I would also claim something else: the principle that every fact in mathematics must have an explanation. This has nothing to do with causality. For example, anything that was purely the result of a calculation was not considered by us to be a good proof. (in Guedj, 1981/1985, p. 22)

Finally, historian of science Catherine Chevalley, writing of her father Claude, memorably observed:

> For him, mathematical rigour consisted of producing a new object which could then become immutable. If you look at the way my father worked, it seems that it was this which counted more than anything, this production of an object which, subsequently, became inert, in short dead. It could no longer be altered or transformed. This was, however, without a single negative connotation. Yet it should be said that my father was probably the only member of Bourbaki who saw mathematics as a means of putting objects to death for aesthetic reasons. (in Chouchan, 1995, pp. 37-38; *my translation*)

This whole extract refers to the themes of this chapter: that of constructing mathematical objects 'well' and then admiring them once they were permanently fixed, if no longer living, no longer warm. [3] After all, surely:

> Truth is truth
> To the end of reckoning.
> (Shakespeare, *Measure for Measure, Act V, Scene 1*)

Nevertheless, all of these mathematicians I have cited above were making broader aesthetic claims in their whys than simply the pursuit of 'the true'.

However, the preserve of the true is not restricted to mathematicians. For instance, Paul Cézanne, in a letter dated 23rd October, 1905 to his friend Emile Bernard (1926, p. 67) signed off with the following acknowledgement and promise:

> I owe you the truth in painting and I will tell it to you.

There is a nice ambiguity here concerning both how to parse the phrase 'truth in painting' [4], as well as why he chose/needed to 'tell it' rather than 'show it', evoking for me Wittgenstein's (1922/1958) striking assertion that "What *can* be shown *cannot* be said" (p. 79; *italics in original*). In an important sense, this chapter is trying to examine possible demarcations between image and word in relation to mathematics, as well as beliefs about the relative propriety of those things that can be shown and those that can be said. Artist Theo van Doesburg (1930/1974) proclaimed:

> The evolution of painting is nothing but an intellectual search for the truth by means of a visual culture. [...] We are painters who think and measure. [...] Most painters work like pastry-cooks and milliners. In contrast we use mathematical data (whether Euclidean or not) and science, that is to say, intellectual means. [...] All instruments that were created by the intellect due to a need for perfection are recommended. (pp. 181-182)

In so doing, van Doesburg was evoking similar concerns to mathematicians, namely attention to truth, to method and to perfection, albeit in a 'visual culture' (see Pimm, 2001). Taken in the context of images, these three foci of attention (linking, respectively, to the why, the how and back to the why again) connect with this chapter's central themes. Discussion related to a human 'need for perfection' is also continued in the closing Chapter ω. In particular, I wish to explore here to what extent elements of twentieth-century *mathematical* culture incorporated and drew on the visual and to what extent they rejected it. Part of this exploration will involve the juxtaposition of views of mathematicians and practitioners of the arts, views that I find at times strikingly parallel.

But before pursuing these threads too assiduously, I feel we need to take heed of van Doesburg's sometime colleague and friend, the artist El Lissitsky (1925/1968), who, using the abbreviation 'A.' for art, warned:

> the parallels between A. and mathematics must be drawn very carefully, for every time they overlap, it is fatal for A. (p. 348)

Making Sense of 'Aesthetics'

Many books on philosophy trace the origin of the word 'aesthetics' back to Alexander Baumgarten (1750/1961), specifically the Latin of his book title *Aesthetica*. However, because of the word's etymological origin in the Greek verb *aisthanomain,* meaning "to perceive", and the noun *aisthesis,* meaning "sensorial perception", the meaning of 'aesthetics' for me is firmly rooted in the senses by means of which we perceive. [5] I hence take the term 'aesthetic' here to refer more to the sensorial than (necessarily) the beautiful and to mathematical objects and artifacts (such as theorems, proofs or methods) equally well as to instances of visual art (such as paintings, collages or sculptures). Additionally, I choose to link the aesthetic to the expression 'ways of seeing' (see, for example, Berger, 1972), as well as to the 'whys' that are offered for seeing things in this way rather than that.

As the chapter title suggests, my central concern here is the question of the drawn (the image and, especially, the diagrammatic) and its complex place in mathematics, not least its movement in and out of favour at various times and in different cultures. This concern cuts across all three of my earlier-identified foci of the what, the how and the why, as images have had a strong involvement in all three.

For instance, Nicolas Goodman (1983), albeit in the context of a discussion of Errett Bishop's constructivist philosophy of mathematics, observed that:

> There has been a strong trend in the development of mathematics in
> the twentieth century to replace seeing with understanding. (p. 63)

He was remarking upon a shift from computational to conceptual proofs, proofs that were frequently non-constructive in nature. (Such arguments seek to convince by method alone rather than by making present.) But his words, which evoke Chevalley's claim about what made a good proof for Bourbaki, also indirectly draw our attention to the move during much of the last century *away* from visual representations in mathematics, *away* from the senses.

The nature of Goodman's claim is complex, as vision provides both a key means of understanding as well as comprising a core metaphor for it in English. (For example, I can say both "I can see *that* it is true" and "I can see *why* it is true".) Vision plays a central role in both mathematics and art (despite some philosophers wishing to disconnect mathematics from 'sense experience').

Sight is also mathematically implicated in the very word 'theorem', with its roots in the Greek verb *theorein* – 'to look at' – and the noun *theoria,* referring to 'contemplation' (especially of *Theos* – God). Thus, vision – seen as a human sense – has important links to the religious metaphor of 'light' and, hence, to the spiritual. As René Thom (1971) noted:

> And according to a long-forgotten etymology a theorem is above
> all the object of a vision. (p. 697)

But what of direct mathematical connections between theorem and image? As one instance, albeit a negative one, James Gleick (1987), in his discussion of the mathematical antecedents of chaos theory, observed:

> In part, Bourbaki began in reaction to Poincaré. [...] Logical analy-
> sis was central. A mathematician had to begin with solid first prin-
> ciples and deduce all the rest from them. The group stressed the
> primacy of mathematics among sciences, and also insisted upon a
> detachment from other sciences. Mathematics was mathematics –
> it could not be valued in terms of its application to real physical
> phenomena. *And above all, Bourbaki rejected the use of pictures.* A
> mathematician could always be fooled by his visual apparatus.
> Geometry was untrustworthy. Mathematics should be pure, formal,
> and austere. (pp. 88-89; *my emphasis*)

Although somewhat over-enthusiastic in tone, Gleick was nevertheless expressing one common view about the work and influence of this significant group of predominantly French mathematicians.

In the next section, I will look more closely at this claim with regard to the shunning of images, with more mathematical voices present. For now, I shall make do with the words of Pierre Cartier (one of the later Bourbaki) who, when asked in interview why there were no images in the Bourbaki books, responded in the following terms: "The Bourbaki were Puritans and Puritans are strongly opposed to pictorial representations of truths of their faith" (in Senechal, 1998, p. 27).

The Loss of the Image

In *Purity and Danger,* her study of pollution and cleanliness rituals (a study of theology interacting with anthropology), Mary Douglas (1966) identified a variety of ways in which 'dirt' offends against order and 'hygiene' offers a set of processes to try to ensure it. There is an identifiable connection between her notion of purity and the mathematician's need to maintain the purity of 'pure' mathematics (and its particular interaction with the unease – or stronger – about how 'pictures' debase this purity).

Douglas singles out classification (a common pure mathematical activity) as a key element underlying rituals by means of which purity is maintained. One question that the term 'pure' mathematics invites concerns the nature of this purity, as well as enquiring of what contagion has it been cleansed, leaving it undefiled. And what dangers stalk, sensed perhaps but unseen, in the long grass?

André Weil (1948), admittedly somewhat in passing, put it thus:

> However, if logic is the hygiene of the mathematician, it does not provide his food; the major problems comprise the daily bread by which he lives. (p. 309; *my translation*)

When Charles Hermite wrote in 1893 (to analyst Thomas Stieltjes right in the middle of a technical letter dated May 20th) in the following terms:

> I turn away in terror and horror from this lamentable plague of functions which have no derivatives. (1893/1905, p. 318; *my translation*)

we sense we are in Douglas's territory of both ritual pollution and taboo. Similarly, in 1908, and also in regard to real functions, Henri Poincaré complained that:

> Logic sometimes begets monsters. For fifty years a host of bizarre functions have been conjured up which seem to be doing their best to be as unlike as possible honest functions which serve some purpose. No longer continuity, or perhaps continuity but no derivatives, etc. What is more, from the logical point of view, it is these strange functions which are the most general; those one runs into without having sought them out no longer show up except as a special case. They are granted only a very small corner.
>
> Formerly, when one came up with a new function, there was a practical end in mind; today, they are expressly invented to show up the reasoning of our forebears and nothing more will ever come of them than that.

> If logic were the sole guide of the teacher [6], one would have to begin with the most general functions, that is to say with the most bizarre. It is the novice who would need to be placed in touch with this teratological museum [...] (pp. 131-132; *my translation*)

Poincaré too was writing in terms that Mary Douglas would recognise and understand. [7]

There is a wonderful symbolic link between 'logical' and 'teratological', between the at-times errant ('monstrous') progeniture of the mind and of the body. This link evokes so well Douglas's chilling example of the Nuer who powerfully reclassify anomalous "monstrous births" as "baby hippopotamuses accidentally born to humans". This conceptual move permits the humane and appropriate response of their immediate return to their rightful place: "They gently lay them in the river where they belong" (p. 39).

Finally, in discussing the dangers of over-abstraction from the empirical, John von Neumann (1947, p. 196) expressed an antithetical concern to that of preserving purity [8]:

> In other words, at a great distance from its empirical source, or after much 'abstract' inbreeding, a mathematical subject is in danger of degeneration.

I wish to claim there is a non-trivial parallel to be drawn between the threat of pollution that Douglas examines in a variety of cultures and the threat of the diagrammatic image tainting mathematics, in particular its 'logic'. The specific threat is that their inclusion would thereby render mathematics' proof rituals ineffective, thereby resulting in a need to declare images taboo.

In the next sub-section, therefore, I explore instances of the image being seen as such a monster in a religious context, as something somehow threatening, of which theology (and pure mathematics) needed to be purified.

The discarded image

C. S. Lewis's (1964) book of this title is about the vanished mediaeval world-view in Europe and how hard it can be to make sense of contemporary literature without knowing what it was. Much of that world-view was religious and imagistic. For instance, in her extensive study of the erasure of images due to English Reformation iconoclasm in the sixteenth and seventeenth centuries, Margaret Aston (1988) wrote on the importance of images:

> The image was not peripheral to medieval Christianity. It was a central means for the individual to establish contact with God. [...] Imagery passed from being a means of instruction to being a means of communication between worshipper and worshipped. (p. 20)

There is a sense, however, in which images themselves (at least those connected to Catholic religious art of the European mediaeval period) were forcibly discarded completely in certain quarters. And mid-twentieth century published mathematics in Western Europe and North America saw a similar discarding of images in advanced textbooks. As quoted in the introduction to this chapter, in response to the question 'Why is there a lack of any kind of visual illustration in most of Bourbaki?', Pierre Cartier commented, "the Bourbaki were Puritans". He went on the expand on his view:

> The number of Protestants and Jews in the Bourbaki group was overwhelming. And you know that the French Protestants especially are very close to Jews in spirit. I have some Jewish background and I was raised a Huguenot [Calvinist protestant]. We are people of the Bible, of the Old Testament [...] (in Senechal, 1998, p. 27)

Cartier's observation provides a helpful link between twentieth-century mathematical presentation practices and a somewhat-forgotten period in the history of Western European religious art. This was especially so in England, where (unlike in France) a form of protestantism became the established faith. In consequence, the iconoclastic effects – because both sanctioned by authority and even institutionalised in law – were the more extreme and consequently the more visible.

In his history of British art, Andrew Graham-Dixon (1996) observes of a reformed English church:

> It is a place that has been purged and purified, denuded of image and relic and every last vestige of superstition. It has been stripped of everything, in fact, except the Word. (p. 32)

I am struck how this quotation could equally well describe tomes of Bourbaki's *Eléments de Mathématique*. This description also echoes the earlier words of a fifteenth-century English archbishop, who mused, "Were it a fair thing to come into a church and see therein none image?" (cited in Aston, 1988, p. 143). Graham-Dixon goes on:

> Their argument [the first English Protestants] with the old religion, the Roman Catholic faith, was that it confused the things of this world with the things of the next [...] They were disturbed by what they saw as a haemorrhaging of holiness from its proper place, heaven, so that, dispersed among the images and cult statues and shrines and relics of the Catholic world, it had become fatally diluted. [...] To the Reformer, the image was quite simply a false idol, the symbol of a dark and demon-haunted pagan past which was soon to be left behind. Only the Bible, the authentic word of God, read and preached and inwardly meditated upon, could lead people to salvation. (p. 35)

How familiar this argument seems when transplanted to mathematics and the role of diagrammatic images there. However, does it make sense to see the Bourbaki as a self-appointed group of Puritans going about their business of effacing, even eradicating, images, whitewashing or smashing them out of existence? I think not. In her book, Aston revives a then-contemporary now-obsolete pair of terms 'iconomach/iconomachy' alongside the more familiar duet of 'iconoclast/iconoclasm' [9]:

> I use *iconomach* alongside *iconoclast* to distinguish those who were
> hostile to religious images from those who wished to break them.
> [...] While iconoclasm necessarily involved iconomachy, the reverse
> was by no means true. Not all opponents were destroyers. (p. 18)

I am not claiming the Bourbaki were iconoclasts in this sense, but the power and influence over professional mathematical practice which they came to wield in parts of western Europe and north America (on this see, for instance, Mandelbrot, 1989) in the three decades following the second world war perhaps enabled their public iconomachy to become institutionalised as a form of iconoclasm.

In mathematics, as much as in religious art, one core issue is that of substitutability: the image can seem as real as the thing, can take its place. In geometry, as Dick Tahta has observed to me, the symbol for a circle is a circle. If for no other reason than its English prepositional collocation, to be an image implies it should be an image *of* something. This, in turn, necessarily requires an image be distinct from the thing ('the original') of which it is an image; images are consequently in a secondary, dependent relation to things. As Aquinas argued, "the image must distinguish itself from the original – if not, it vanishes into it" (in Besançon, 2000, p. 163). [10] To have an image is to fail to have the thing. (In passing, the same is true of 'painting' – to be a painting immediately gives rise to the question 'what is it a painting of?', a question that became an initial challenge for 'abstract' art. While this is also true of words, it seems a far less common move to demand what this word is a symbol for.)

> The historical relations between sacred image, nature and truth,
> distinct from any theoretical history of knowledge or ideas, persist-
> ently attempt to describe, explain or reconcile the distance between
> icon or image and an 'authenticity' of artistic figuration. (Steiner,
> 1992, p. 7)

Or, as Graham-Dixon (1996) observed about church iconography, "Images were signposts to the next world, placed in this one" (p. 18).

One concern of the Bourbaki was their insistence on the separation between symbol and object, and the rooting out of *abus de notation,* despite the fact such confusions between 'the things of this world and the things of the next' can actually significantly improve mathematical performance. Mathematics relies on links between surface symbolic forms and mathematical ideas, mathematical 'objects', on being able to move reliably from representation to 'truth'. One of the challenges in exploring the place of images in mathematics involves questions of how this is to be done.

Aston once again:

> When the iconoclasts went to work they were concerned with atti-
> tudes as well as objects. They wanted to erase not only the idols
> defiling God's churches, but also the idols infecting people's
> thoughts. [...] The faith was remade by what believers were shut off
> from, as well as by the new certainties they bumped into. (p. 2)

One thing that Bourbaki readers were shut off from was images (another was specific examples, a third applications). And as the influence and power of this group and its publications grew in the post-world-war-II period, this style of mathematical 'worship' also grew. Aston goes on, emphasising the systematic and the ideological nature of the iconoclast enterprise, while also seemingly prefiguring both the frame and scope of the Bourbaki project:

> Protestant iconoclasts believed that widespread destruction was necessary for the renewal of an entire religious system. [...] They regarded themselves, indeed, as having taken on a task comparable to the first conversion of the world. (pp. 5, 9)

When Pierre Cartier said of Bourbaki that:

> The stated goal of Bourbaki was to create a *new mathematics*. He didn't cite any other mathematical texts. Bourbaki is self-sufficient. [...] It was the time of ideology. Bourbaki was to be the New Euclid, he would write a textbook for the next 2000 years. (p. 27)

the parallel seems strikingly strong. The Bourbaki even referred to the collection of already-published volumes in their encyclopaedic *Eléments de Mathématique* as the 'canonical Bible' (see, for example, Beaulieu, 2000, p. 227). And François Le Lionnais (1962), in the strongly pro-Bourbaki collection he edited *Great Currents of Mathematical Thought,* entitled the first part of the expanded two-volume second edition, 'The temple of mathematics'.

While describing the core of the twentieth century as a 'century' of ideology, Pierre Cartier also specifically compared the manifestos of the surrealists and the Italian futurists in art with the Introduction of Bourbaki's first 'fascicle' of results. [11] In mathematics, that ideology was in large part one based on method. But it also arose from a de-emphasising of particular content, of reframing the core element of mathematical attention as that of relation rather than object.

Finally, as a transition to questions of 'method', in his introduction to the first part, 'Structures', of this collection, Le Lionnais (1948) lauded:

> Mr Nicolas Bourbaki, that many-headed mathematician, who has undertaken to reformulate the exposition of Mathematics by taking it from its starting points – not historical but logical – and then striving to reconstruct the complexity by means of materials passed through the sieve of the axiomatic critique. (p. 22; *my translation*)

Modern Madness in Method

> Greater than the temptations of beauty are those of method.
> (James Richardson, 2001, p. 26)

The method in focus here, unsurprisingly, is the axiomatic method. But it has a curious history. And despite Bourbaki's insistence on mathematics as a singular, unitary entity, he often referred to the axiomatic method in the plural, as 'axiomatic methods'. For instance, in a call to arms found in an article entitled 'Modern axiomatic methods', Dieudonné (1962/1971) exhorted us to follow him:

> Hence the absolute necessity from now on for every mathematician concerned with intellectual probity to present his reasonings in *axiomatic* form, i.e., in a form where propositions are linked *by virtue of rules of logic only,* all intuitive 'evidence' which may suggest expressions to the mind being deliberately disregarded. (p. 253)

Logic (and its 'cold light') and intuition (to which images become linked) are thus polarised, with intuition and its attendant images vilified for mathematical *presentation* as opposed to discovery.

One customary place to start in discussions of 'modern' mathematics is with B. L. van der Waerden's influential 1930 treatise *Moderne Algebra*. (It was only with its fourth edition in 1959 that the adjective 'modern' was dropped, reflecting both a chronological shift of reference, but also that the orientation to algebra it offered had become mainstream.) This text, as well as the work of Hilbert, Noether and other German mathematicians, formed the main exemplars that inspired the Bourbaki group.

A full and detailed examination of this text, not least in terms of its public inauguration of a full embodiment of 'modern' algebra, can be found in Corry (1996). Corry attempts a retrospective characterisation of the 'structural' approach in algebra and includes the axiomatic method as a key necessary but not sufficient condition. For him, this approach also de-emphasises solvability of equations and the particularities of the real or complex numbers. These contributory elements are motivated by certain aesthetic valuing of generality and abstraction, for instance, over computability (first with equations, later with, for example, Galois groups).

Just before the first publication of van der Waerden's book, the prominent German algebraist Helmut Hasse (1930/1986) gave a public lecture at the Deutsche Mathematiker-Vereinigung entitled 'The modern algebraic method'. For him, this algebraic method helped to discern and even to create a likeness among unlike things – they were similar because they could be treated as if they were the same:

> Modern algebra winds a unifying band of method around essentially different things and in this way contributes to the required organic and systematic unification of mathematics. (p. 19)

Hasse characterised this method both in terms of contrast with those of analysis and with earlier algebraic approaches. He offered two reasons for operating in an abstract setting: the greatest possible generality of content and the greatest possible economy of means. It is this latter aim that "represents the key difference between the older and the modern conceptions of algebra" (p. 19).

Embedded in a discussion of the appropriate location of the "so-called fundamental theorem of algebra" (p. 20), Hasse drew attention to subject-matter frontiers:

> As things stand now, what must be said when the idle question is
> posed of what is the boundary between algebra and the others
> mathematical disciplines, is that this is not so much a question of
> *substance* but rather of *method*. (p. 20; *italics in original*)

In this view of the far greater salience in *modern* algebra of the *how* over
the *what,* Hasse significantly predated the trenchant expression of the core
of Modernist painting by its chief advocate, the critic Clement Greenberg
(1960/1965):

> The essence of Modernism lies, as I see it, in the use of the char-
> acteristic methods of a discipline to criticize the discipline itself –
> not in order to subvert it, but to entrench it more firmly in its area
> of competence. [...] It quickly emerged that the unique and proper
> area of competence of each art coincided with all that was unique
> to the nature of its medium. The task of self-criticism became to
> eliminate from the effects of each art any and every effect that
> might conceivably be borrowed from or by the medium of any
> other art. Thereby each art would be rendered 'pure', and in its
> 'purity' find the guarantee of its standards of quality as well as of
> its independence. 'Purity' meant self-definition, and the enterprise
> of self-criticism in the arts became one of self-definition with a
> vengeance. (pp. 193-194)

Thus, in regard to Greenberg's first sentence above, we can see the extensive growth of 'meta-mathematics' during the first half of the twentieth century as being just such a self-critical examination, one using the tools of
mathematics and logic to explore the scope of mathematics itself. In particular, this reflects what Corry (1996, 1997) refers to as a 'reflexive' use of
mathematics, in order to ascertain its own methodological limits, another
aesthetic criterion. This development also connects to the query raised in
Chapter α concerning possible reasons for the rise of meta-mathematical
activity and preoccupation at the beginning of the twentieth century. But
considerably earlier, Greenberg (1939) had written:

> The very values in the name of which he [the poet or artist]
> invokes the absolute are relative values, the values of aesthetics.
> And so he turns out to be imitating, not God – and here I use 'imi-
> tate' in its Aristotelian sense – but the disciplines and processes or
> art and literature themselves. This is the genesis of the 'abstract'. In
> turning his attention away from subject-matter or common experi-
> ence, the poet or artist turns it in upon the medium of his own
> craft. The non-representational or 'abstract', if it is to have aesthetic
> validity, cannot be arbitrary and accidental, but must stem from
> obedience to some worthy constraint or original. This constraint,
> once the world of common, extraverted experience has been
> renounced, can only be found in the very processes or disciplines
> by which art and literature have already imitated the former. These
> themselves become the subject matter of art and literature. If, to
> continue with Aristotle, all art and literature are imitation, then
> what we have here is the imitation of imitat*ing*. (pp. 36-37)

Shading one's eyes only slightly, it is possible to read this as a description of modern algebra, of a broad method becoming the object of study once the material world has faded (with applied mathematics seen as a form of imitation). And the guide that saves the mathematician from a study of the arbitrary is an aesthetic one. This view is echoed, in part, in von Neumann's (1947) chapter 'The mathematician', where towards the end he repeatedly touched on the role of 'the aesthetic', offering two criteria for salvation from mathematics becoming:

> more and more purely aestheticizing, more and more purely *l'art pour l'art*. This need not be bad, if the field is surrounded by correlated subjects, which still have closer empirical connections, or if the discipline is under the influence of men with an exceptionally well-developed taste. (p. 196)

When republished in *The Mathematical Intelligencer*, Hasse's essay was accompanied by some 'Comments' made by Bruce Chandler and Wihelm Magnus (1986) some sixty years later, in which they revisionistically claimed:

> What Hasse calls the 'modern algebraic method' is actually the axiomatic method. [...] It is strange that he claims the method as being typical for algebra. The axiomatic method started, of course, with geometry. (p. 24)

For them, in other words, the 'proper area of competence' of 'the axiomatic method' was the whole of mathematics. Yet Hasse deliberately addressed his comments "to analysts whose approach and methods are directly opposite to modern algebraic techniques" (p. 18). For me, the key point to underline is Hasse's claim that this method did originate both in the twentieth century and in algebra, a subject area that has frequently struggled in terms of its objects. This is perhaps because algebra is more concerned with the scope of its techniques, its computations and manipulations, and gets increasingly detached from specific objects. [12] In the Bourbaki movement, we see an extensive implementation of 'the' axiomatic method (which, like its self-styled scientific forebear, seems only subsequently to attract the assertive definitiveness of the definite article) put forward as *the* arbiter (judge, authority, touchstone) of 'good' mathematics or even the 'only' mathematics, the one true faith.

Time and Motion Studies

> The Cubists preferred to be intelligent, scientific, and said "It's not enough to see, one must also know". From this moment, the constructive arts are seen to be completely new foundations. The content of such works is no longer aestheticism or painting as such, but the two moments of time and movement. From my point of view, our epoch is the dynamic epoch. [...] Now there is simply movement. (Kazimir Malevich, in Gavin, 1990)

In cubism, we find representations of plurality, of elements of atomised seeing occurring at the same time, as if the viewer were able to occupy different
points of view at one time, or as if the viewer were seeing through the eyes
of several observers at once.

> Picasso had pictured the fifth *Demoiselle* from two diametrically
> opposed points of view simultaneously, which not only exploded
> the Renaissance convention, but also undermined a whole series
> of the conventions by which we represent simultaneity in an art
> object that doesn't move. (Everdell, 1997, p. 249)

Cubism was about plurality of point of view, sharing different perspectives
'at once'. Futurism was more concerned with machines in motion, providing
static images of the dynamic moving object, not only from multiple points
of view but also at plural times. Both of these early twentieth-century art
movements involved further undermining classical conventions of much of
the previous centuries.

In historical academic painting, there was a strong aesthetic of not allowing
brushmarks to be visible in finished paintings. (This was so instilled in the
community as to be seen as a mark of competence – it is frequently the fate
of avant-garde art to be criticised as incompetent. [13]) The means of production were not to be in evidence, nor was the meaner, the one who meant it
to be made. The drawer was thereby 'encouraged' to participate in his or her
own erasure (apart from the only text customarily found on the image, the
painter's signature), along with any sign of the process of drawing the drawn.
Along with the person, most traces of time have been effaced.

Kandinsky (1926/1979) wrote:

> Time, in painting, is a question in itself and is very complex. [...]
> The apparently clear and justifiable division:
>
>> Painting – space (plane)
>> Music – time
>
> has upon closer, though yet hasty, examination suddenly become
> doubtful and, as far as I know, this first became apparent to the
> painters. The tendency to overlook the time element in painting
> today still persists. (pp. 34-35)

Mathematics is overwhelmingly concerned with generality, with discovering
and deploying ways of dealing with the many as if it were one – and doing
so at one and the same time. The previous section has alluded to the
increasing decontextualisation of mathematics during the twentieth century
with the march of 'modern' algebraic methods, in particular in the Bourbaki
push for 'maximal generality' (see Corry, 1997).

For over two thousand years, public mathematical text has been attended
by two further absentings: detemporalisation and depersonalisation. (This
important 'de-' triad is discussed further in Chapter ω, and can be found in
Balacheff, 1988.) Both detemporalisation and depersonalisation connect into
my third category mentioned at the outset, the *why*, which is related to

questions of 'the aesthetic of purity' in pure mathematics. My central focus in this section, then, is on certain systematic means by which *time* is erased from mathematical writing, while exploring why it necessarily lingers on with mathematical diagrams (which thereby suggests a further possible reason for mathematical iconomachy).

In the next section, I look at the role of the *person* in mathematical text, by means of the notion of 'agency'. For, in the words of Preston Hammer's aphorism, the most neglected existence theorem in mathematics is the existence of people (in Brookes, 1970, p. vii). However, as will soon become apparent, detemporalisation and depersonalisation are very closely linked.

On time in mathematics

> As it was in the beginning, is now, and ever shall be, world without
> end. Amen. (part of the Christian lesser doxology)

This ancient text – part biblical ('world without end'), part early addition in the fifth century AD – evokes for me the mathematician's world, one without time, where truths are always and already true. What are some contributors to how this sense is achieved? Below, I offer four different sources, two arising from mathematical text practices, two related to the nature of geometric diagrams and the complex procedures that emerge from the need to link them to the written text. The latter pair are, however, only two among the many 'brushstrokes' in the sand that need assiduously to be erased.

(a) Some temporal textual practices

One contribution is undoubtedly through the imposed syntax of mathematical prose itself, in particular by means of its customary verb tense structure and other sequencing markers. (Part of learning mathematics is learning to speak and, especially, to write mathematically.) One significant feature of mathematical text is the extensive use of connectives, such as *hence, therefore, but,* words which explicitly mark relationships among antecedent and subsequent sentences and clauses. Most of the connectives used in mathematical English, such as *then, hence, since* and *when,* as well as the key terms *ever, always* and *never,* also have a chronological sense in everyday English.

Solomon and O'Neill (1998), for instance, examined diverse types of writing by mathematician William Rowan Hamilton about his discovery of quaternions, not just in published papers, but also in more personal communication and correspondence. They report on contrasting structures among the various types of writing:

> An examination of the letter and the notebook reveals a more complex structure than a simple narrative. The texts contain two distinct component texts: a *mathematical* text is embedded within a *personal narrative*. The difference between the texts is indicated in the tense system, the choice of deictic reference and the forms of textual cohesion employed. (p. 216; *italics in original*)

Solomon and O'Neill go on to observe that the personal narrative is written in the past tense and it also contains deictic time markers like 'yesterday' [14], whereas the mathematical material is:

> in the timeless present. At the same time there is also in the math-
> ematical sub-text a distinct form of cohesion: the temporal order in
> the narrative gives way to a logical order in the mathematics. [...]
> mathematics cannot be narrative for it is structured around logical
> and not temporal relations. (pp. 216-217)

Thus, quite simply put, mathematics comes to be seen as a-temporal because that is how it is 'supposed' to be written about and how it is done by prefessionals. In school and even university, students learn that, "Last night I found that this happened" is historicised narrative, whereas shifting first to "you find" or "one finds" and then to the stripped Euclidean asser-tion "this is always the case" brings rewards. [15] (I return to this in the next section on agency.)

A second, quite different, way in which the human past is brought into the present in contemporary mathematics is by the widespread use of eponyms, namely the naming of concepts, results, theorems and principles by means of mathematicians' names. According to Henwood and Rival (1979), Charles Darwin complained about this practice in biology more than a century ago on two grounds: inciting hasty, and therefore shallow, work (in order to ensure the patriarchal priority of surname), as well as the impen-etrability of mere names over descriptive naming. Mathematical examples of the latter would be 'Abelian' rather than, say, 'commutative' groups and 'Hilbert' rather than, say, 'complete normed vector space over the complex numbers, where the norm is given by an inner product'. In mathematics, in response to Darwin's former complaint, such naming often is done by others: in Hilbert's case, by Stone and von Neumann (see Rota, 1997, p. 199). Arguably, this is due to the reification that comes with abstraction.

I wish to add another 'complaint' here: the posthumous co-opting of the past by the present, in order to render mathematical styles apparently less temporal. By labelling a particular formulation as Lagrange's or Cauchy's theorem, for example, is to recruit their work into a modern idiom, into 'our' style of doing mathematics. I believe it is done, in part, to make them one of 'us'. [16]

In this way, the 'timelessness' of the notions, formulations and even styles of proof themselves become buttressed by the past and, thus, ren-dered invisible. What we lose (and are meant to lose) is a contingent and historical sense of our present-day mathematics. How can someone nowa-days read Cayley or Jordan, having studied a course in group theory which included versions of 'Cayley's theorem' and the 'Jordan-Hölder theorem', and not assume that they were addressing the same objects, which were seen in the same axiomatically-specified way. [17]

(b) Under construction: drawings being drawn in time

In the Ancient Greek world, there was considerable discussion of the 'time-lessness' of geometry and the permanenence of geometric knowledge versus the *genesis* (coming-into-being) of mathematical objects by means of con-structions. Lachterman (1989) wrote:

> For Speusippus the language of *genesis* has an 'as-if' character. […]
> we must not treat the constructions and motions on display in a
> geometrical proof as 'makings' in the course of actual performance,
> as time-consuming just for the reason that first this is done, then
> afterwards that is done, and so forth. […] What, taken literally,
> seems now to be coming into being for the first time […] must be
> regarded figuratively as having already been accomplished all
> along. (p. 62; *italics in original*)

Lachterman quoted Aristotle, in *On the Heavens,* arguing against a parallel between diagrammatic proofs and the cosmic myth put forward in Plato's *Timaeus* [I 10 280a4ff]:

> They say that ordered things came to be out of disordered, but it is
> impossible for the same to be simultaneously disordered and
> ordered. There must be a genesis involving the separation [of things]
> in time as well. In the diagrams, nothing is separated in/by time.
> (p. 63)

Lacheterman concluded this section (p. 65) with an examination of the curi-ous verb tense and mood of Euclidean construction language of 'operations' – the perfect passive imperative: examples include 'Let such-and-such have been done' and 'let it have come about that …'.

> The perfect tense tells us that the relevant operation has already
> been executed prior to the reader's encounter with the unfolding
> proof. […] Euclid invites us, not to perform the operation on our
> own, nor to observe him performing the operation before our
> eyes, but rather to consider the operation as already anonymously
> [the agentless passive mood] performed before the 'present
> moment' […] This verbal operator does not so much suppress time
> as shift it backwards into an unnoticed past […] [18]

In mathematics, much is made of assertions by *fiat*, that is sentences imper-atively beginning 'Let …' But who is the suppressed person toward whom such an imperative is directed? At least with 'Let us pray' there is an implied audience or congregation. But are all 'Let …' utterances implied, inclusive, first-person plurals? Or are they appeals for permission to a higher authority? Whence derives the authority to make such utterances?

In the English translation of Euclid, such fiats are, almost entirely, simply uttered, unattributed to any speaker: only occasionally will someone declare 'I say that […]', followed by the specific statement of the result *in terms of the letters in the diagram.* The earlier *protasis* ("enunciation") is usually simply asserted, without reference to any diagram or an asserter. (For an instance of this rigid structure for setting out a proposition, see Netz, 1999, pp. 10-11.)

It is worth recalling that the Word preceded the Light and was used to call it into being: the most familiar word to follow *fiat* is *'lux'*. In the Bible, God's utterances are preceded by an attribution tag: 'And God said' comes before 'Let there be light'. In the Genesis account of creation, God spoke first to call into being (the ultimate in speech acts) and then, subsequently, judged the effect by sight, 'And saw that it was good': vision and detachment.

The above makes reference to the fact of letters being used to glue diagram to text. Netz (1998) explores principles underlying Greek use of 'baptism' of diagrams by letters and argues convincingly for the diagram and not the text being the central Greek mathematical object (see also the final section of this chapter). He also makes a strong case for the baptism of a geometric object by alphabetic letters being both predominantly in alphabetic order and compact (in the sense that no 'intermediate' letters are omitted). In addition, labelling of points that are completely undetermined by the text were alphabetical had they been determined at the appropriate moment in the course of the proof. The Greek principles of labelling carry with them sequential information about emergent construction.

In other words, one further locale where traces of time ordering are seemingly evoked occurs in labelled diagrams by means of their alphabetic lettering. This provides a human path laying down letters through the image, though to the wary this may be seen to defile it. Netz's work has provided us with a compelling account of what he terms the 'archaeology' of the mathematical diagram in Ancient Greek mathematics.

Moving mathematics

In true Wittgenstinian fashion, in *Republic* (527A), Plato has Socrates attempt to undermine elements that might suggest motion and time in geometry, specifically action, as indicators of human presence.

> Their [geometers'] language is most ludicrous, though they cannot help it, for they speak as if they were doing something and as if all their words were directed towards action. For all their talk is of squaring and adding and applying and the like, whereas in fact the real object of the entire study is pure knowledge. (in Molland, 1991, p. 182)

In parallel with the hiding of brushstrokes suggesting a timeless and agent-less 'work of art', geometric construction lines are also to be made to disappear, once their task has been carried out. As mentioned above, in Euclid, for instance, constructions are achieved in no 'proof'-time whatsoever by means of a particular evocation, the perfect passive imperative. In *The Geometer's Sketchpad* (discussed in the previous chapter), Hide Line is a universal but reversible imperative feature. However, neither appearance nor disappearance is permanent, the lines are both there and not there at the same time. And the infinitely repeatable Undo command allows travel back in time, to reveal precisely the linear diagrammatic history that accompanies every *Sketchpad* 'figure' – in some important sense, Sketchpad figures are temporal text objects.

Tom Wolfe (1975/2002), in his lively polemic *The Painted Word,* reported himself astonished to realise:

> Modern Art has become completely literary: the paintings and other works exist only to illustrate the text. (p. 4; italics in original)

The 'modern' diagram never loses its temporality: it is forever a literary object. When I look at a *Sketchpad* image, an instance of a modern geometry of a sort, I see a voice (as Bottom masquerading as Pyramus in *A Midsummer Night's Dream* would have it). [19] The voice is reciting a script, the voice is talking, coaxing, persuading the image into existence, a voice without which the image cannot be summoned, without which it has no identity. I can, with the software's help, travel backwards in time. I can revisit the drawing (as it was drawn in the beginning, as it is being drawn now and as it ever shall be so drawn) to hear and see once again the tale of its genesis, a tale that is forever being told.

Mathematical Agency

> But there is quite another way of thinking about science [other than 'representation']. One can start from the idea that the world is filled not, in the first instance, with facts and observations, but with *agency*. The world, I want to say, is continually *doing things,* things that bear upon us not as observation statements upon disembodied intellects but as forces upon material beings.
> (Andrew Pickering, 1995, p. 6; *italics in original*)

The notion of agency is not one that has been widely discussed in relation to mathematics. Pickering's account of modern science centrally involves the interaction of 'human' and what he terms 'material' agency. Pickering's view of scientists is as "human agents in a field of material agency which they struggle to capture in machines" (p. 21). This dialectic of resistance and accommodation – performers engaged in 'a dance of agency' – he terms the 'mangle' of practice: underneath lie questions of action, cause and efficacy. After giving an example of the weather, he adds:

> Much of everyday life, I would say, has this character of coping with material agency, agency that comes at us from outside the human realm and that cannot be reduced to anything within that realm. (p. 6)

In mathematics, at first sight at least, things seem different. What is and where might lie the mathematical parallel to material (non-human) agency?

Equally, although on the other side, there are sterling efforts made to shed any suggestion of human agency at work in written mathematics. In her book *Writing Mathematically,* Candia Morgan (1998, Chapter 2) examines a published mathematical paper (taken from the *Journal of the London*

Mathematical Society) in terms of its syntactic and pragmatic features. Many relate to the absence of the author. I summarise:

- *distant authorial voice* (e.g. use of passive, absence of direct author or reader references);
- *extensive use of nominalisations* (transforming processes and actions into objects: e.g. 'stabilizer', 'permutation', 'discriminant' or creating more complex noun phrases);
- *the use of imperatives rather than pronouns* (e.g. directives to 'consider', 'suppose', 'define', 'let') – though this does indirectly suggest a human presence.

The removal of the presence of the mathematical author by genre conventions of either using 'we' or extensive passive constructions interacts with the perennial present tense (discussed in the previous section). Additionally, the anthropomorphising of mathematical objects – so they are animated to have apparent agency of their own – contributes to their gaining a permanent, 'timeless' existence. One important effect of these nominalisations is the obscuring of human agency: grammatically, the possibility is thereby generated that nominal expressions can themselves become actors, that is active subject-position participants in the text, a *cause* of other phenomena. This, I suggest, is one significant means by which a parallel to material agency is created.

Pickering does attempt to discuss agency in the development of mathematical concepts. He offers the notion of the agency of the discipline. In an extended example of William Rowan Hamilton's work that culminated in his construction of the quaternions on October 16th, 1843, Pickering finds places of choice and discretion ("the classic attributes of human agency", p. 116). He also finds others, "where the disciplinary agency [...] carries scientists along, where scientists become passive in the face of their training and established procedures" (p. 116).

Mathematicians are not simply free to create, despite some grandiose (even child-like omnipotent) statements to the contrary. But the fact that they are not only constrained either, not simply passive observers in the face of a pre-existent (pre-ordained?) mathematical realm, that there are free as well as forced moves or choices, is one place where the possibility of an aesthetic dimension to mathematics arises. As John von Neumann (1947) observed:

> And in all these fields [those that are furthest from their empirical, human roots, including modern, "abstract" algebra and topology] the mathematician's subjective criterion of success, of the worthwhileness of his effort, is very much self-contained and aesthetical and free (or nearly free) of empirical connections. [...] I think that it is correct to say that his criteria of selection, and also those of success, are mainly aesthetical. (pp. 191, 194)

In particular, such choices (both selecting and justifying) arose for the Bourbaki with regard to axiom systems. Corry (1997) has posed the following challenge:

> Dieudonné did not hesitate to use the term 'axiomatic trash' (1982, 620), to designate theories based upon the axiomatic treatment of systems that he considered unimportant or uninteresting. But what actually is the criterion for winnowing the chaff of 'axiomatic trash' from the wheat of the mathematically significant axiomatic systems? (p. 279)

However, Corry's article also documents (albeit not in those terms) the considerable lengths Bourbaki went to in order to *mask* human agency in this regard, in order to bolster their strongly promoted claims for the 'eternal' nature of their work. There is an interesting connection between Bourbaki's hiding of human values and choices and the more general curious 'hiddenness' of the aesthetic in mathematics.

One aspect of mathematics that continually interests me is the way in which certain disciplinary practices serve to efface or even erase the human mathematical agent, to remove one side of Pickering's mangle, thus contributing to the apparent inevitability, permanence and necessarily 'independent' existence of mathematical objects and methods. The other is to acknowledge that human mathematical choices are not made in a vacuum, that past 'callings-into-being' also influence what is possible in the present.

Pickering's example is essentially algebraic, the increasingly dominant, verbally orientated, nineteenth- and twentieth-century mathematical worldview. In the next two sub-sections, I wish to touch briefly on the way in which different forms of 'disciplinary agency' seemingly stem from words and images.

The agency of the letter

> Every sign *by itself* seems dead. *What* gives it life? (Wittgenstein, 1953/1963, p. 128; *italics in original*)

In contemporary mathematics, unlike in ancient Greek mathematics, the letter is algebraic; it is essentially a 'saying' (whether spoken or written), linear in time, of one thing after another. It is used as if it were an empty symbol (to lighten the load for ease of manipulation, for calculation); it offers itself up as a substitute, a counterpart for action; it hides its object. Dieudonné (1962/1971) claimed:

> But the difficulty [of differing intuition and imagery, of problems of infinity, of the essence of a proposition being its *content*] vanishes if on the contrary one agrees that the essence of a proposition is its *form,* in other words, if one agrees that it is needless for a proposition to evoke any other mental image than the perception of the symbols used to write it. (p. 260; *italics in original*)

This, for Bourbaki, corresponded to the defining issue in modern art of 'flatness', 'the integrity of the picture plane' and the 'essence of the surface' from Clement Greenberg's characterisation of American Abstract Expressionism in the 1940s and 1950s. The content is to disappear, the symbols no longer represent: they are to become the main event, symbol becomes object. 'Purity', for both groups, becomes the watchword.

Mathematical argument, mathematical text is claimed to have a certain disembodied agency, not perhaps in the material world but in the mathematical realm. It purportedly 'shows' things *must* be this way; they *cannot* be otherwise. When a result is first proven, does it conceivably *make* it so, whereas before it was only potentially so?

A textual mathematical proof is a certain kind of act, one whose agency and efficacy depends on it being properly conducted. One aspect of this is the 'principle of retransmission of falsity': make a mistake somewhere in the proof and the proof should not work. In the speech act language of Austin (1962) and Searle (1969), a proof is a 'performative', it is supposed to do something in being 'uttered'. (Netz's conception of an ancient Greek mathematician addressing a diagram orally fits this sense well.) More specifically, just as Cézanne's statement to Emile Bernard was an overt *promise* to provide 'the truth in painting', and to say it, so simply uttering the word 'proof' at the top of a mathematical text is to make an implied promise to an implied reader.

For a proof to 'work', it must be correctly 'uttered', invoked. For instance, simply shuffle the order of its sentences and it has an efficacy comparable with a similarly-scrambled marriage ceremony: the couple are not, in fact, married, the theorem is not, in fact, proven. (In passing, this is another temporal aspect a proof retains.) To use Wittgenstein's term, such a purported proof 'misfires'. [20]

Mary Douglas (1966), whose work on pollution and taboo I mentioned earlier, described the universe of a 'primitive' world-view as "personal": that is, one that is responsive to signs and symbols.

> The most obvious example of impersonal powers being thought responsive to symbolic communication is the belief in sorcery. The sorcerer is the magician who tries to transform the path of events by symbolic enactment. He may use gestures or plain words in spells or incantations. Now words are the proper mode of communication between persons. If there is an idea that words correctly said are essential to the efficacy of an action, then, although the thing spoken to cannot answer back, there is a belief in a limited kind of one-way verbal communication. And this belief obscures the clear thing-status of the thing being addressed. (p. 86)

Recall Chapter α introduced a number of historical instances of the magical and the mathematical overlapping (in as possibly a worrying manner as the concern of Lissitsky mentioned at the outset of this chapter). The truth of the Word is supposedly independent of its asserter: we give assent to the assertion not the asserter. And yet, once again, there is mathematical resonance for me in Douglas's description: a proof arguably places constraints on the mathematical world, it exerts a form of agency.

The agency of the image

In mathematics, the image is geometric; it is essentially a 'seeing' (see note [19] for more on the mathematical consequence of 'seeing' versus 'saying'). It is at least two-dimensional, hence not easily mapped onto a time sequence (though its drawing-in-time can often produce one temporarily). An image has structure, but no necessary grammar, no necessarily 'right' way to be read. [21]

The diagram is inanimate, yet is active in the proof process. Netz (1998, 1999) attributes it central agency in the Greek proving process. He has mathematicians dressing diagrams (in letters) and then addressing them, much as the sorcerer does in the Douglas passage cited above. It raises for me the following two questions:

> What do we *ask* a mathematical diagram?
> What do we *ask of* a diagram in mathematics?

Derrida (1978/1987, p. 4) has provided the comparable 'performative' question about images, when he asked, "Does speech act theory have its counterpart in painting?" For me, the question is in what sense does an image or diagram possess or carry disciplinary agency? (Recall, too, the remarks of Nicholas Jackiw in Chapter 7 on ego and agency in relation to *Sketchpad* diagrams.) There is no room in this chapter to explore these questions, but I feel them to be of considerable importance in understanding what the loss from Bourbaki's iconomachy actually was.

Mathematics and Its Objects

We have in some sense come full circle. I started this chapter asking about *what* mathematicians attend to, *how* they attend and *why*. From a deliberate twentieth-century eschewing of diagrams as a necessary or even allowable object of attention, the methods of 'modern' algebra have been employed to create new objects of attention. Ironically, perhaps, the 'whys' have remained pretty much the same. This brief, concluding section returns to the question of 'what is attended to?'.

Whether written or drawn, intermediary symbols are fundamentally employed in the practice of mathematics. Both in their different ways cease simply to 'represent'. In the absence of a 'true' object, like a cuckoo's egg hatching in a nest, they subvert, supplant and replace, becoming instead the object of attention. In each instance, consequently, both the algebraic letter and the geometric diagram then revert to being icons in the traditional, religious sense: that is, to recall Graham-Dixon's words, "signposts to the next world, placed in this one".

According to Besançon (2000), an icon:

> is an instrument of contemplation [*theoria*] through which the soul breaks free of the sensible world and enters the world of divine illumination. (p. 134)

Or, as Simone Weil (1952) offered:

> Method for understanding images, symbols, etc. Not to try to inter-
> pret them, but to look at them until the light suddenly dawns.
> Generally speaking, a method for the exercise of the intelligence
> by means of looking. (p. 109)

Netz (1998) has a very clear view of what might, in sharp contrast to the
Hilbertian version, be called 'Greek formalism'. He claims that, for the
ancient Greeks, quite unlike for us now when the object is defined by
means of verbal axiomatic systems, the object of mathematics was the diagram: "All the signification is contained in the diagram and the text is parasitic on this signification" (p. 38). My starting concern was a sense that the
image was looking pale and drawn in mathematics, and of late had been
withdrawn from many texts. This is a far cry from Archimedes boasting of
the extent of his publication by the number of drawings (*diagrammata* –
equally 'diagram' or 'proposition' – see Netz, 1999).

Just as with 'instrumental' music, which originally accompanied and
enhanced text (whether sacred or secular, choral chant or popular ballad)
before being presented on its own, algebra did not initially have to worry
about its objects. For when referents and reference were needed, it referred to
geometric ones: algebra was an embellishment, an ornament or enhancement.
Subsequently (with Euler, say), 'classical' objects of algebraic awareness and
attention, such as 'equation' or 'polynomial in one or more variables', were the
surface text objects that had been generated during algebraic activity. What the
modern algebraic method finally allowed was the very creation of objects to
which its methods applied, by fiat. It reflected an aesthetic, one which in a
curious but important sense valued actions or operations, *ways* of proceeding, over things: yet it proceeded by turning them into further things in need
of new methods.

In conclusion, in keeping with the opening quotation from Lebesgue, I
see this chapter as part of a study 'on the foundations of mathematics and
mathematical method', but – following Kandinsky – not one that has solely
relied on mathematics.

> Words conjure
> what is called.
>
> [...]
>
> What we name
> grows.
>
> (Nelson, 2002, p. 23)

Acknowledgements

One important starting point for this chapter which I gratefully acknowledge was a protracted series of agreeable disagreements more than a decade ago with artist and art teacher Birute Macijauskas, concerning whether mathematics or art were the more abstract, the more aesthetic, the more spiritual, the more human endeavour. I am also highly indebted to Dick Tahta, both for inspiration and for direct assistance with this chapter. Lastly, I wish to pay tribute to the considerable indirect influence on this chapter of the late David Fowler, who was one of my most important teachers. [22]

Notes

[1] This quotation brings to mind Walter Benjamin's phrase: 'the disinterested intentionlessness of truth'. John Fauvel (1988), in his article on the marked differences between the Cartesian and Euclidean rhetorical styles, expressed the latter's attitude toward the reader as follows:

> Euclid's attitude is perfectly straightforward: there is no sign that he notices the existence of readers at all. [...] The reader is never addressed. (p. 25)

It is as if Euclid is turning his back on the audience, as jazz trumpeter Miles Davis used to do on occasion. But Euclid's *Elements* are certainly far from intentionless as a whole – they are organised/structured to various purposes, even if each proposition by itself is simply, and apparently disinterestedly, asserted and proven (QED), each construction-problem unassumedly resolved (QEF).

[2] According to Proclus at least, Hippocrates was the first compiler of mathematical 'Elements'. The Bourbaki, in their monumental set of texts produced some 2400 years later, not coincidentally entitled *Eléments de Mathématique,* aspired to be the last. But whereas the ancient Greek notion and techniques of proof necessarily and essentially involved lettered diagrams – see Knorr (1975) or Netz (1999) – any diagrams (apart from the later, internally-contested use of commutative 'diagrams' and concomitant 'diagram-chasing' arguments) were completely eschewed by Bourbaki. Thus, while claiming to echo and evoke the Euclidean project of *Elements* in their own, the Bourbaki renounced its central methodological means, namely diagrams.

For instance, the very opening words of the *Introduction* to the first book of the first part of Bourbaki (*Théorie des Ensembles*) declare:

> Ever since the Greeks, whomsoever says mathematic[s] says proof; some even doubt whether proof is to be found, outside of mathematics, in the precise and rigorous sense that this word gained from the Greeks and which will be given it here. It is justifiable to claim this meaning has not changed, because what was a proof for Euclid is still one to our eyes. (1960a, p. 1; *my translation*)

(In French, 'mathematics' is both plural and attracts a plural verb. Bourbaki claimed there was only one 'mathematic': I would add, 'theirs'.)

[3] In that same interview (Guedj, 1981/1985), Claude Chevalley said, "It [a bible in mathematics] is a very well arranged cemetery with a beautiful array of tombstones" (p. 20). In his analysis of Jackson Pollock's paintings, Robert Steiner (1992) comments on:

> the 'self-portrait' statement in Vermeer's *Allegory of Painting*, in which the face is known only by a death mask obliquely seen on the table beside the embodiment of Clio, muse of history – in effect, an ego at once present and not [...] (p. 78)

This seems fitting to juxtapose against the Chevalley quotations, in part because of the open question of where traces of the Bourbaki ego might lie in the singularly self-styled *Eléments de mathématique*. But Steiner's comment also seems apposite because of the presence of Clio, the historian's muse and marker of temporal presence. For not only did Bourbaki (1960b) shape a curiously self-serving history of mathematics in their own image, but it was also Clio's followers who formed part of the forces who have helped revive some mathematics that Claude Chevalley and the other Bourbaki believed they had put to the sword – or at least had tried to provide with a decent and enduring burial. See, for instance, 'The withering immortality of Nicolas Bourbaki' (Aubin, 1997).

[4] Jacques Derrida (1978/1987), in a characteristically complex and frequently perplexing work entitled *The Truth in Painting*, takes Cézanne's statement as his starting point for an extensive examination of aesthetics and painting. To what extent does it matter whether the truth is to be *told* (as in the English moral injunction to children to do just that) or to be *shown* ('revealed')? As William Blake optimistically claimed: "Truth can never be told so as to be understood, and not be believ'd".

[5] The human senses which have little if any mathematical import are those of taste and smell. Yet 'taste', along with 'judgement', is one of the central metaphors of aesthetics, as well as one of the hardest notions to discuss. Consider, for example, modernist art critic Clement Greenberg's (1978/1999) observation that "when no esthetic value judgment, no verdict of taste is there, then art isn't there either" (p. 62). There is also an important distinction in the Greek meaning between the *aesthetic* and the *noetic,* namely between the sensible (or perceivable) and the reasonable (or conceivable). This very old *epistemological* distinction also sheds some additional light on the title of this book, in that a far more 'conventional' title, in terms of how mathematics is customarily seen, would have been *Mathematics and the Noetic*. But part of what the actual title and sub-title is asserting is mathematics' connection with an ancient perceived affinity with the senses. (For much more, including further connections between art and mathematics, see the feisty book by Robert Dixon (1995) entitled *The Baumgarten Corruption*.)

[6] In addition, Poincaré was once more anticipating the future, in that Bourbaki texts were written on the explicit principle of 'from the most general to the particular' (Chevalley, in Guedj, 1981/1985, p. 20) and unwitting students were indeed placed in touch with them. Claude Chevalley once again:

> There was something which repelled us all: everything we wrote would be useless for teaching. (p. 20)

However, there is no room here to explore this pedagogic thread: for more on this, see Love and Pimm (1996).

[7] The language of monsters, not least as immortalised in his evocative term 'monster-barring', is invoked in Imre Lakatos's (1976) book *Proofs and Refutations: the Logic of Mathematical Discovery*. This is a historical–mathematical study of nineteenth-century mathematicians' practices surrounding both Euler's 'conjecture' for polyhedra V − E + F = 2 and the 'discovery' of uniform continuity.

[8] On the question of the dangers of 'inbreeding', David Antin (1987) has challengingly written:

> the viability of a genre [here, perhaps, a mathematical style of presentation] like the viability of a family is based on survival, and the indispensable property of a surviving family is a continuing ability to take in new members who bring fresh genetic material into the old reservoir. So the viability of a genre may depend fairly heavily on an avant-garde activity that has often been seen as threatening its very existence, but is more accurately seen as opening its present to its past and to its future. (p. 479)

It is unclear to me to what extent Bourbaki's work can reasonably be seen as 'avant-garde activity', not least in terms of it precisely attempting to cut mathematics off from its past and to freeze the future.

[9] Without using the term *iconomach*, Bruno Latour (2002) makes a contemporary attempt to enrich the categorisation of 'iconoclastic gestures' towards images. He differentiates five types (by means of labels A to E). As are against all images; Bs are against 'freeze-framing', not against images; Cs are not against images, except those of their opponents; Ds break images unwittingly; Es are simply the people, mockers of iconoclasts and iconophiles alike.

> What distinguishes the As from all other types of iconoclasts is that they believe it is not only necessary but also possible to *entirely* dispose of intermediaries and to access truth, objectivity, and sanctity. [...] Between images and symbols you have to choose or be damned. Type A is thus the pure form of 'classical' iconoclasm, recognizable in the formalist's rejection of imagination, drawing, and models [...] Purification is their goal. (p. 26)

Jean Dieudonné, in the preface to his 1969 textbook *Linear Algebra and Geometry*, wrote that he omitted all diagrams on purpose, "only to show that they are unnecessary" (p. 13) to a development of linear algebra and geometry. But this was quite a while after the Bourbaki revolution, long after the complete rejection of geometric imagery in favour of an algebra- and text-based aesthetic had been achieved.

With regard to Bs, according to Latour, what they object to is the singling out of a single image from the plenitude.

> What they [Bs] fight is *freeze-framing*, that is, extracting an image out of the flow, and becoming fascinated by it, as if it were sufficient, as if all movement had stopped. What they are after is not a world free of images, purified of all the obstacles, rid of all mediators, but on the contrary, a world *filled* with active images, moving mediators. (pp. 26-27)

The previous chapter in this book on *The Geometer's Sketchpad* gives an instance of software for mathematical Bs. Can any one diagram act as a general image or is movement endlessly circulating between particulars? Is even speaking of a 'general image' necessarily a catachresis or even an oxymoron? Latour offers the early twentieth-century Russian artist Kazimir Malevich as an archetypal type B iconoclast.

[10] Frayn (1974) notes:

> When our symbols *resemble* what they represent, we teeter on the brink of nullity, on the edge of the Duchamps [*sic*] tautology, the urinal labelled 'Urinal'. Only the act of selection, of framing, of labelling, differentiates what is represented from what represents it. The challenge to our reading instinct here is one of minimality. (para. 155; *italics in original*)

The whole mathematical challenge of geometric diagrams is captured here in this single observation. The 'framing' and 'labelling' of lettered diagrams comprise a core, contested component of Greek mathematics (see Netz, 1999).

[11] Mumford (1991) has commented:

> The 20th century has been, until recently, an era of 'modern mathematics' in a sense quite parallel to 'modern art' or 'modern architecture' or 'modern music.' That is to say, it turned to an analysis of abstraction, it glorified purity and tried to simplify its results until the roots of each idea were manifest. (p. xxvii)

Later in this chapter, I draw the parallel with modern art a little more explicitly.

[12] Unguru (1994) has written about Jacob Klein's (1968) distinction between 'the generality of the method' and 'the generality of the object' of investigation in Greek. mathematics:

> According to Klein, Greek mathematical methods are dictated by the ontology of the *mathematiká,* the mathematical objects, while modern mathematics starts with a general method and is led by it to the features of the mathematical objects. (p. 214)

[13] This holds true in mathematics also. Recall, for example, the monster-barring comment about Hilbert's purported (non-constructive existence) proof of a generalisation of a result of Paul Gordan's, who had earlier established a finite basis for binary forms of any degree. It was Gordan himself who scathingly remarked, "Das ist nicht Mathematik, das ist Theologie" (in Reid, 1986, p. 34). Of course, one of the intents of this and the next chapter is to explore the possibility of a considerable overlap between arguments in mathematics and in theology. And, according to Morris Kline (1972), Gordan did also revise his position, subsequently observing that, "I have convinced myself that theology also has its advantages" (p. 930).

[14] 'Deixis' or 'deictic forms' (e.g. you, now, here, this, that, there) comprise elements of language that 'point' to particular surroundings (referents such as persons, times, places), to the *context* of the utterance. The origin of the word is closely related to the term *deiknume,* which refers to a direct 'showing' form of mathematical proof, one of the earliest known proof types. 'Deixis' is also a component of the word *apodeixis,* the Euclidean term for the actual demonstration part of any proof of a proposition.

[15] Netz (2000) makes an astute observation about a possible but not actual temporal and voice difference between Greek proofs by *analysis* and those by *synthesis:*

> Even without any second-order pronouncements [i.e. meta-com-mentary about purpose or intention], there could have been sug-gestions of a sequence of discovery, e.g. using a past tense in the assertions of the analysis as opposed to the present tense in the assertions of the synthesis, or using a first person active for the constructions in the analysis as opposed to the third person passive for the constructions in the synthesis. But nothing like this happens, everything is in the present tense or the third person passive sug-gesting the impersonal work of mathematical necessity rather than the accident of authentic discovery. (p. 146)

[16] This practice is quite reminiscent of the Mormon belief that baptising one's ancestors in the present can bring them retroactively into the church. Interestingly, no mathematical result or procedure bears the name 'Bourbaki', despite the refer-ence having assumed some of the desired 'depersonalised, detemporalised' features of mathematics itself (though it is still a male label). At a meta-level, there are per-haps good reasons to call the 'axiomatic method' the 'Hilbert method' – certainly the Bourbaki thought so, carrying out their program in his name. However, there are some interesting and important differences between Hilbert's and Bourbaki's own views of this method, particularly concerning the nature of axioms – see Corry (1997).

[17] In his novel *Small World*, David Lodge (1984) had a character, Persse McGar-rigle, claim to have done a thesis "about the influence of T. S. Eliot on Shakespeare" (p. 51). When another character queries this, Persse replies:

> we can't avoid reading Shakespeare through the lens of T. S. Eilot's poetry. I mean, who can read *Hamlet* today without thinking of 'Prufrock'? Who can hear the speeches of Ferdinand in *The Tempest* without being reminded of 'The Fire Sermon' section of *The Waste Land?* (p. 52)

But at least we do not call it Shakespeare-Eliot's *Hamlet*.

[18] Linguist Stephen Levinson (1983) distinguished between L-tense and M-tense (linguistic and meta-linguistic tense, respectively). The former is what we normally think of as 'grammatical' tense and the latter has to do with time reference *within* an account. The M-tense structure is usually organised with respect to some sort of 'coding time' (usually an event within the account, with respect to which events are marked as being prior to, contemporaneous with or later than). M-tense and L-tense attributions may coincide or differ. Gerofsky (1996, 2004), for example, has exam-ined verbally-posed mathematics problems from this point of view.

Although not cast into the linguistic language of Levinson's M-tense and coding time, Lachterman's discussion of Euclidean discursive practice fits it perfectly. I can-not explore this further here. However, I am interested in the issue of how M-tense functions within mathematical proofs in general. In particular, I plan to explore else-where the question of what the coding time is for a mathematical proof

[19] A contributory source for this chapter was a series of discussions with Dick Tahta (author of the next chapter in this book), about connections and contrasts between the senses of hearing and sight in relation to mathematics (see Tahta and Pimm, 2001; for more on the visual versus the tactile, see Ivins, 1946). This exchange was triggered by a mutual reading of Jonathan Rée's (1999) book *I See a Voice: a Philosophical History of Language, Deafness and the Senses*. The 'seeing of voices' of Rée's title concerns work with deaf children who are not nowadays discouraged from 'signing'. Among other speculations, Tahta raised the possibility of cardinal number being predominantly visual while ordinal number is predominantly aural/oral. But what we had not yet examined was the marked extent to which being rooted in different human senses could lead to different aesthetics, particularly in relation to mathematics.

Rée gives us a pair of extreme examples from Greek mythology, those of Narcissus and Echo, with regard to how people encounter the world. The former dwelt so much in his eyes, he fell in love with his own image; while the latter was so much in her ears, she heard nothing but her own voice. "She [Echo] was a mouth as well as an ear, but he [Narcissus] was only an eye" (p. 71). This chapter, in part, starts to examine the relative mathematical dominance of mouth-and-ear over eye, of Echo over Narcissus, in the twentieth-century mathematical realm.

Netz (1998, 1999) also explores this oral nature of Greek mathematics, seen as a silent dress rehearsal of an argument in front of a lettered diagram, prior to its final writing down (in order to be sent to a necessarily distant mathematician – he claims "necessarily" as they were so thin on the ground). In Greek mathematics, I would say, Narcissus and Echo were both strongly in evidence.

[20] James Joyce's verbal playfulness was at full force in *Finnegan's Wake*:

> My unchanging Word is sacred. [...] Till Breath us depart! [...] The ring man in the rong shop but the rite words by the rote order! (1939/1959, p. 167)

In the context under discussion here, perhaps 'the right words by the wrote order' might be more apt.

[21] These two quotations below connect to questions of grammar.

> Grammar tells what kind of object anything is.
> (Wittgenstein, 1953/1963, p. 116)

> Strangely enough, a grammar in art today still seems ominously dangerous to many.
> (Kandinsky, 1926/1979, p. 84)

And William Ivins (1969) wrote:

> Thus while there is very definitely a syntax in the putting together, the making, of visual images, once they are put together there is no syntax for the reading of their meaning. With rare exceptions, we see a picture first as a whole, and only after having seen it as a whole do we analyse it into its component parts. [...] This leads me to wonder whether the constantly recurring philosophical discussion as to which comes first, the parts or the whole, is not merely a derivative of the different syntactical situations exemplified on the one hand by visual statements and on the other by the

necessary arrangement of word symbols in a time order. Thus it may be that the points and lines of geometry are not things at all but merely syntactical dodges. (pp. 61-62)

For more on this, see Pimm (1995).

[22] In Pimm (2004), I have attempted to explore some aspects of a fine teacher's influence, in relation to David Fowler. His extensive work on the history of Greek mathematics, as well as his way of being in the world, both as an academic and as a human being, proved profoundly shaping for me. I dedicate this chapter to his memory.

Figure 1: Christ Pantocrator, fourteenth-century mosaic, St Saviour in Chora, Istanbul

CHAPTER 9
Sensible Objects

Dick Tahta

The most imposing icon in early eighth-century Byzantium was probably the mosaic image of Christ above the Chalke, the bronze gate entrance to the palace built by Justinian to the south of Sancta Sophia. It is said to have been similar in style to a surviving fourteenth-century mosaic (Figure 1, left) in the restored church of St Saviour in Chora. [1]

By the 720s, however, there was a growing hostility – at any rate among the ruling élite – to icons of any sort and a succession of edicts by the emperor Leo III led to the gradual destruction of religious images all over the city. In, or soon after, 726, he ordered his soldiers to replace the Chalke mosaic with a cross. According to some accounts, this so incensed a group of devout women that they assaulted and killed the officer in charge. Their ringleader, St Theodosia, was seized by the soldiers and summarily executed. [2] Demonstrations against the removal of icons were followed by more stringent measures (including the removal of the cross that had replaced the Chalke mosaic) and the so-called Iconoclasts controlled the city for the next hundred years or so.

It may be supposed that the spread of Islam, a religion which took the Mosaic injunction against graven images very seriously, had an influence on the developing unease at the widespread veneration of holy relics and icons. There was also a continuing theological controversy within the Church itself about the nature of Christ. Those who tended to emphasise his human nature deplored the way his image could be granted divine qualities. It seemed to the emperor and his advisors that Christianity had lost its particular vision and had sunk into superstition. It is often also the case that religious disputes cloak power struggles; and secular rulers have always found some recompense in the dissolving of monasteries and the sacking of churches.

Both parties in the ensuing struggles were, in fact, agreed about one particular icon, namely the Eucharist. The image in this case was an established church ritual which, it was believed, involved the divine presence. But this did not involve depiction of any sort. Moses had heard God's voice, but saw no form: God was, in the theological jargon, 'uncircumscribable'. His presence in the Eucharist was achieved by a process known as 'economy' to theologians. This is perhaps more simply explained in terms of synecdoche, the figure of speech in which the whole is signified by its part. God is held to be partly manifest in the sacraments, which then 'stand for' him figuratively – but, it was also supposed, completely.

Where the Iconoclasts disagreed was when it came to the same view being held about images of Christ. Since God was uncircumscribable, he could not be depicted. Hence, ran the iconoclast argument, any representation of Christ could only depict his human nature and so should not be venerated in terms of a divine presence. The argument was even stronger in the case of the innumerable and increasingly popular icons of the Virgin Mary or the Saints.

Sense and Sensibility

The reader may at this stage be wondering where this historical digression is leading. Theological disputes seem far removed from mathematics or its possible links with aesthetics. There were, of course, some mathematical issues behind the theological disagreements. For example, the distinction between the relations of 'same' and 'similar' was at the heart of the often-bitter polemics about the nature of Christ. [3] But here the relevance will be – despite the Iconoclasts – that icons do have a significance that I hope to invoke when considering the nature of other objects.

The iconoclast controversy raised the issue of how an image is related to whatever it is supposed to re-present – that which was classically called its 'prototype'. For St Theodore [4], a staunch defender of icons, image and prototype were of the same category. He invoked Aristotle's example of a similarly linked pair, the double and the half.

> For the prototype always implies the image of which it is the prototype, and the double always implies the half in relation to which it is called double. For there would not be a prototype if there were no image; there would not even be any double, if some half were not understood. But since these things exist simultaneously, they are understood and subsist together. (1981, p. 110)

It was not the image itself that was venerated, but the 'form' of the prototype which is 'in-corporated' into the image. And, according to Theodore, "they have their being in each other". Christ was held to be the prototype of his own image, but here the part signified the whole in more than a figurative sense. For Theodore, a linked definition meant linked simultaneous existence. Christ's image did not imitate, or re-semble, but rather *partook in* the being of Christ. To be like was to be. Theodore described an icon as a "self-manifested vision", almost like moonlight, which *is* sunlight rather than an image of it. Miguel Tamen (2001) comments [5]:

> In sum, the extension of the term "Christ" includes "sensory appearance of Christ," which is to say that, much to our Western theological horror, there is a very important sense in which an image of Christ *is* already Christ, and a very important sense in which Christ *is* already an image of Christ. (p. 24; *italics in original*)

It is said that when the mosaic image of Christ was removed from the Chalke, Leo caused a poem to be attached which deplored the speechless nature of the icon. Of course, we might now suppose that whether an image 'speaks' to us is a matter of whether and how we listen. And then, indeed, what it is we hear. A similar distinction between our inner and outer worlds may be that which lies between looking at an icon, or indeed any object, and what in fact it is we see.

A baby lies in the cot and seems to be looking at what an observer might say was its hand. But what is seen is as-yet inchoate. Eyes fasten on the strange object floating in front of them; the baby does not yet know what the hand does, what it is for, let alone what it is called. Michael Frayn (1974, paras 287-288) once suggested that to see something is to make a private metaphor of it:

> The metaphor might be a visual one: I see the shifting vapour as a face, I see my hand as a hand [...] The likeness of your hand to a hand has got trodden down into the subsoil of your perception just as the metaphors for mental states have got trodden down into your thinking about yourself.

Sensation will eventually be linked with reflection: *sensible* (i.e. perceptible) objects become *sensible* (i.e. comprehensible). Inner and outer are inextricably linked: as the saying goes, the innocent eye is blind and the virgin mind is empty. Moreover, it seems that eye and mind do, after all, contain much that is uncircumscribable: whatever you look at can never be fully expressed, nor can what you see. And so, people soon make different sense of sensible objects.

Reason and Imagination

A dramatic example of this – in Western thought at any rate – is the so-called 'dissociation of sensibility' in the seventeenth century. Until the rise of science as we now know it, reason and imagination were woven into the same fabric. More literally, one might say, scratched in the same clay. For, in ancient Babylonian cuneiform writing, the great god Anu was written with one mark of the wedge – and this mark also stood for the number 1 (or 60). The scribe would have recorded myths *and* charted the heavens. The bard would have celebrated heroes *and* described nature. Those who told tales were also the ones who took tallies.

But, inevitably, as distinctions were made, boundaries were delineated. In fact, writing and reckoning may nowadays seem hostile to each other. A famous illustration of this occurred in the account of a dinner party at the end of 1817, given by the painter Benjamin Haydon for various literary friends, including Wordsworth, Lamb and Keats. The party agreed that Newton had destroyed all the poetry of the rainbow in explaining its colours. They drank to the toast, "Newton's health and confusion to mathematics!". [6]

Keats wrote about the rainbow a few years later. In a passage in his poem, *Lamia,* he referred scathingly to 'cold philosophy', by which he meant mathematics and science (called 'natural philosophy' in his time).

> There was an awful rainbow once in heaven:
> We know her woof, her texture; she is given
> In the dull catalogue of common things.
> Philosophy will clip an Angel's wings,
> Conquer all mysteries by rule and line,
> Empty the haunted air, and gnomed mine –
> Unweave a rainbow, [...]

For the poet, the rainbow may be a miraculous symbol of hope: "My heart leaps up when I behold a rainbow in the sky", wrote Wordsworth. In the Old Testament, of course, the rainbow was a token of God's covenant with Noah and succeeding generations that there would be no more catastrophic floods (Figure 2). In other accounts, the promise of the rainbow is seen to be elusive. You may dream of the crock of gold at the end of the rainbow, but in reality this always moves on just when you think you are getting there.

Figure 2:   Noah, his family, and animals leaving the Ark, thirteenth-century mosaic, San Marco, Venice (courtesy of Dumbarton Oaks)

The rainbow is a delicate subjective vision that differs for each observer. It is perhaps this aspect of the rainbow that first stirs the curiosity of those who would know why and how it is as it is. It is interesting to note the many mathematicians who have tackled such questions from the ancient Greeks onwards. The prosaic style in which Descartes, for one, described his work on the rainbow might only confirm the prejudice of the poets (including that of the later seventeenth-century poet Jean-Baptiste Rousseau, for whom Descartes *"a coupé la gorge à la poésie"*).

> I took my pen and made an accurate calculation of the paths of the rays which fall on the different points of a globe of water [...] and then I found that after one reflection and two refractions there are many more rays which can be seen at an angle of from forty-one to forty-two degrees than at any smaller angle; and that there are none which can be seen at a larger angle. (in Boyer, 1959, pp. 211-212)

But those who can imaginatively enter into a geometric exploration of the shape and size of a rainbow may also find that the heart can 'leap up' at this way of seeing things.

Does the magic of a rainbow reside in the object or in the mind of the observer? Even if we reject the implied duality, there still seem to be different ways of saying what you see. It may be helpful to think of the rainbow (Keats's 'awe-full' thing) as an icon and to see in the different ways of interpreting it echoes of the iconoclast controversy. [7]

Figure 3: Islamic interlacing design, fourteenth-century mural, Alhambra, Granada

Harmony and Proportion

Another example of a supposed dissociation of sensibility might be found in the different ways people might now see the fabulous Islamic decorations of the Alhambra palace in Granada. By at least the fourteenth century, Islamic craftsmen had perfected a style of abstract art that was meant to offer a mystical sense of harmony and unity in the world. In contemplating the complex interlacing patterns of some of the Alhambra tilings, the eye has no reason to pause anywhere; moreover, the flow comes back on itself, so that there is no start or end. Such interlacing was felt to be a direct expression of the idea of the divine unity behind the enormous multiplicity of the world. Harmony was held to be both 'unity in multiplicity' and 'multiplicity in unity' – and interlacing expressed both these aspects (Figure 3). There is an interesting echo of such metaphysical ideas in Georg Cantor's mathematical treatment of infinity, which was, for him, a matter of seeing the Many as a One. [8]

More generally, Islamic designs may, like mathematics itself, offer the observer the freedom of a 'hollow symbol', one that is not tied to 'concrete' reality. So that there are, inevitably, some very particular mathematical interpretations of the Alhambra tilings. For instance, it has been claimed (though this is disputed) that they include an example of each of the seventeen so-called plane crystallographic groups. A mathematical discussion of these would, of course, be for some to 'unweave a rainbow' – as would, to take another example, the following excerpt from a technical account (Grünbaum and Shephard, 1993) of interlacing:

> The number of crossings in a translational repeating unit of a design with group *p4m* corresponding to a 1-strand pattern is $2c(S)$. If the group is *p6m* the number of crossings is $3c(S)$. If there is more than one strand, we replace $c(S)$ in these formulae by the sum of the crossing numbers for each of the strands in the fundamental region. (p. 153)

This sort of interpretation, we may assume, is far removed from whatever were the original intentions of the Islamic craftsmen or their patrons. What these were is not always clear: they probably included some more worldly ambitions as well as spiritual intent.

There would have been similar issues in the flowering of Christian art in Italy from about the same time. To take a fully developed, example from the fifteenth century, consider Piero della Francesca's painting, *Baptism of Christ* (Figure 4).

This was originally the central panel of an altarpiece in a church in Piero's native town, Borgo San Sepolcro. It stood on the altar presiding over the Eucharist and would therefore have had some religious significance at the time. But we do also know that there were some other considerations in the mind of the artist. In his own words:

Painting consists of three principal parts, which we call *disegno, commensuratio* and *colorare*. By *disegno* we mean profiles and contours which enclose objects. By *commensuratio* we mean the profiles and contours set in their proper places in proportion. By *colorare* we mean how colours show themselves on objects [...] (in Baxandall, 1985, p. 112)

Proportion and perspective were key elements in Piero's paintings. Modern critics emphasise the mathematical abstraction that seems to secularise the religious theme.

Figure 4: Piero della Francesca, The Baptism of Christ, *panel, c.1440–1450, photograph © The National Gallery, London*

> Yet even in the Baptism, the least rigidly mathematical of all his
> paintings, we are at once conscious of a geometric framework; and
> a few seconds' analysis shows us that it is divided into thirds, hor-
> izontally, and into quarters vertically. [...] These divisions form a
> central square, which is again divided into thirds and quarters, and
> a triangle drawn within this square, having its apex at the Dove
> and its base at the lower horizontal, gives the central motive of the
> design. (Clark, 1951, p. 13)

Commensuratio revealed a mathematically ordained universe.

Idolatry and Iconoclasm

What may be called the 'mathematisation' of objects, like an Islamic design
or a Christian painting that have had some original iconic significance, may
be said to idolise them in a different way. The Alhambra is a tourist attraction and its tilings are enjoyed, for most viewers, aesthetically rather than reli-
giously. Altarpieces are displayed in galleries and admired for their artistic
and historical interest. Indeed, it might be said that the very notion of a work
of art – a relatively recent notion – is a form of non-destructive iconoclasm.

The justification for the preservation of objects in museums and in
galleries, or archeological sites, is usually couched in terms of their being
saved from neglect or destruction. But, from another point of view, much
of their original significance has thereby been lost. Contemplation of a painting in a gallery, however ecstatic it might be, is not like the veneration of
icons, at any rate as described by an iconophil like St Theodore. The preserved object represents, stands for, in some way. In psychoanalytic jargon,
it is a part-object, but in its original setting its iconic significance is that of a
whole-object (with the figurative link now being a general metonymy).

When the Louvre became a national museum in 1793, it was considered
unfortunate that various objects of some national historical interest also had
undesirable political overtones. Should a sceptre from a desecrated royal
tomb be preserved in the museum? The Monuments Commission decreed it
was not to be considered a sceptre, but rather "an example of fourteenth-
century goldsmith work". As Stanley Idzerda (1954, p. 26), in an article on
iconoclasm in the French Revolution, commented, "Immure a political symbol
in a museum and it becomes merely art".

Museums can be mausoleums. Of course, art lovers would disagree;
they may not have the specific awesome feelings invoked originally by an
object such as a royal sceptre, but they are nevertheless idolaters, even if of
a different persuasion. The Alhambra tilings have different meanings for different groups of people – such as devout Muslims or devoted mathematicians. Not everyone agreed with St Theodore that objects have intentions
and interpretations of their own. Nor has everyone agreed with Locke that
experience is derived solely from what he called 'reflection on sensations' –
the traditional empirical epistemology.

Tamen (2001) suggests that things become interpretable only in the context of what he calls "a society of friends". Friendship is taken here simply to involve the measure of agreement indispensable to any community. He quotes Aristotle: "Friendship would seem to hold cities together". Tamen (p. 3) sees some similarity in the different ways that people share some preoccupation:

> there are no interpretable objects or intentional objects, only what counts as an interpretable object or, better, groups of people for whom certain objects count as interpretable and who, accordingly, deal with certain objects in recognizable ways. Even if there appear to be many, if not always so formally constituted, kinds of such groups, I submit that what allows us to speak of the existence of such societies is roughly the empirical resemblance between what certain people do in relation to tea leaves, and what certain (other or the same) people do in relation to cold fronts, novels, or statues.

Or, we might add, what certain people do to sceptres, altarpieces and tiles or to patterns and symmetries in general. The silent child in Vachel Lindsay's poem, *Euclid,* watches the solemn greybeards drawing circles in the sand. The child watches them "from morning until noon because they drew such charming round pictures of the moon".

Interpretable Objects

So we become, in Tamen's phrase, friends of interpretable objects – whether as religious believers, art-lovers, antique collectors, gardeners, mathematicians, whatever. The objects in such cases are *anathemata* – votive offerings, devoted objects, which, for fourth-century St John Chrysostom, were "laid up from other things". For the poet David Jones (1952, p. 29), they were the signs of something *other,* things which are "set up, lifted up, or in whatever manner made over to the gods". [9] It is intriguing to note that the related word *anathema* originally meant something venerable, but then became restricted to the opposite meaning – the Church now 'anathematises' heresy.

Objects can become special to us, they may become *anathemata;* but they may also become fetishes – part-objects rather than whole ones. Wholes may contain something holy(!) or we may, through them, be contained by something unresolved in us. The ambiguity is emphasised by the reading of the word 'object' as a noun (*ob*-ject) or as a verb (*ob-ject*). The aggressive-sounding word suggests both resistance against our wishes (e.g. an obstacle) and an aim or target (the word's original sense) for our strivings (e.g. an objective). The psychoanalyst Michael Balint (1968) pointed to a further issue in some cases. [10]

> Any threat of being separated from [the object] creates intense anxiety and the most frequently used defence against it is clinging. On the other hand, the object [...] becomes so important that no con-

cern or consideration can be given to it, it must have no separate
interests from the individual's, it must simply be there and, in fact,
it is taken for granted. (p. 69)

It should be noted that 'object' is now being used in a very general sense.
"Mental Things are alone Real", wrote William Blake. When psychoanalysts
refer to objects, they are not only referring to physical things, but may also
be referring to memories, images, dreams, emotions or concepts, as well as
(most confusingly) to people. Anything, in fact, that one might be aware of
or pay attention to.

What do mathematicians attend to? Do they note their own thought
processes? Or are they reading the physical world? Or perhaps noticing a
world of 'forms', something like Wordsworth's "independent world erected
out of pure intelligence"? Practitioners emphasise some or all of these possibilities. And there will be similar variation in their views of what a 'mathematical object' might be. They might well all agree that when attending,
say, to a geometrical diagram, they are not attending to visual marks as such,
but to some abstracted generality that can only be 'seen' by the mind's eye.
This is likely what St Theodore meant when he asserted that generalities are
seen with the mind and thought. And this is also the sense in which many
might agree with Plato that the power of mathematics is to draw the soul
away from the sensible to the intelligible, though already the precise interpretation of these latter words is problematic. [11]

A simple example may illustrate the point. How do we read this diagram?
David Pimm (1995, p. 57) has suggested that we "see *through* the particularity
of [a] diagram to grasp the generality of what the drawer is attempting to
focus attention on". The diagram, then, may be said to act symbolically, it is
"not the object that the theorem is speaking about". [12]

Whatever the associations, consider what happens when one asks for
the area of the outlined shape? In this case, there are seven steps to the
'staircase'. Or could it have been any number? The mathematician may read
a generality in the diagram, by seeing it in terms of half an enclosing square
together with a certain number of small half-squares (those on the 'steps' of
the staircase). [13] The corresponding algebraic object (i.e. the externalisa-
tion of this mental awareness in terms of written signs) would then be the
'expression' $(\frac{1}{2}.n^2 + n.\frac{1}{2})$.

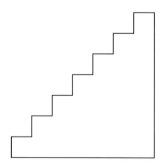

The process here is not unlike the way we have now learned to read paintings. In the pictorial sense, Piero's *Baptism* is a representation of an episode in the life of Christ. But there are far more important symbolic senses. For the religious, the painting represents a miraculous event, promising the possibility of general transfiguration. And, as has already been shown, for the mathematically minded, it may be a matter of harmony and proportion – abstractions that perhaps offer another sort of transfiguration.

"The touch between the acute angle of a triangle and a circle is no less significant", claimed the painter Wassily Kandinsky (1931, p. 352), "than that between the finger of Michelangelo's Adam and God's". We may well ask how mathematical objects come to bear so much investment. One clue, it will be suggested, lies in some of the earliest objects of our experience.

Figure 5:   Pablo Picasso, Fruit and wineglass, *1908, private collection,* © Picasso Estate (Paris)/ SODRAC (Montreal) 2005

Here and There

A two-year-old boy picks up any small object lying to hand and throws it across the room. As he does so, he calls out something that his mother interprets as an announcement that the object has gone away. This becomes a game which he manifestly delights in repeating over and over again. Then it happens one day that what he picks up is a wooden reel attached to a piece of string. He holds on to the free end of the string and throws the reel away. This is accompanied as usual with the sound that he always makes as the object disappears. But now he pulls the reel back with the string and greets its reappearance with a joyful call that seems to be announcing its return. And this game of hide and seek, of rejection and return, is then also repeated over and over again.

Many parents will recognise something similar in their own experience. But they may not agree with the interpretation that is already embedded in this particular account. Mothers are usually in no doubt that they know what baby-talk is actually saying. It is clear to them when their child is joyful or sad. But outsiders are sometimes more cautious, certainly when it comes to any further interpretation of such familiar nursery events: for example, the one that was offered by Sigmund Freud in 1920.

The boy was, in fact, Freud's grandson: the sounds he made were understood by his mother as being the German *fort* (meaning 'gone') and *da* (meaning 'there'). But Freud (1920/1955) of course, then made much more far-reaching interpretations of the game:

> It was related to the child's great cultural achievement – the instinctual renunciation (that is the renunciation of instinctual satisfaction) which he had made in allowing his mother to go away without protesting. He compensated himself for this, as it were, by himself staging the disappearance and return of the objects within his reach [...] her departure had to be enacted as a necessary preliminary to her joyful return, and that it was in the latter that lay the true purpose of the game. (pp. 15-16)

Freud went on to add that the experience was initially passive, but that by repeating the game the boy took on an active role. In throwing away the object, he might, Freud suggested, be satisfying a repressed impulse to punish his mother for going away from him. The game established his control over conflicting emotions. At first, all his toys seemed to be sent away – 'gone'. At a later stage, they could be retrieved and the mastery in being able to do this also contained the pleasure in achieving independence of the mother who had hitherto returned the toys herself. Freud noted further that the pleasure also involved that of conquering the pain of the loss that lurked within the independence. Now the infant could expel and return the mother in his mind: *fort–da* !

Later, of course, repeatedly throwing and retrieving can become even more virtual – writing and reading, adding and subtracting, differentiating and integrating, whatever – but the pleasurable invariant may be preserved. Moreover, as psychoanalysts have observed, a repetition of needs may be accompanied by a need for repetition.

Some readers may have already decided that this is already too far-fetched, that wooden reels are wooden reels and not mothers, that Freudian interpretations are speculative and unprovable. And of course they would be right. However, the issue is, on the one hand, whether inanimate objects can represent (symbolise) people or feelings and, on the other, whether unprovable interpretations may, in fact, turn out to be useful in some way. According to Goethe, *was fruchtbar ist, allein ist wahr* ("only that which is fruitful is true"). This goes against a traditional mathematical grain, but it is the approach being taken here. Thus, one initial postulate (a Euclidean *aitema* or demand) is that objects of any sort are not simply what they phenomenologically seem, they can 'stand for' other things. And that what these latter are – and whether they are held to be conscious or unconscious – is a matter of interpretation, which is to be understood as a possibly fruitful construction rather than a provable truth. [14]

Discovery and Creation

A range of experiences can be condensed into single objects, images, words or sounds. And much of this process may be unconscious, in the sense that it is not easily accessible, sometimes deeply repressed. This means that, what is being triggered at an unconscious level is often quite different from – and perhaps contradictory to – what is being considered consciously. This can be confirmed in our experience of painting, literature or music.

It must also be part of our experience of mathematics. Thus, consider proof by *reductio ad absurdum,* where the crunch point is when you encounter a contradiction: A cannot be not-A. The 'cannot' here is very much a conscious prohibition. But, according to psychoanalysts, the usual rules of logic do not hold for unconscious processes. Freud famously said there was no 'not' in the unconscious. Thus, at some level, I both hate and love at the same time, the wooden reel is and is not my mother.

One analyst, Ignacio Matte Blanco (1975), has developed this further in asserting that there is no order in the unconscious – no before or after, no larger or smaller: in technical terms, there are no asymmetric relations. So that, in unconscious 'thought', the icon of Christ is Christ, Piero's painting is the Baptism and the breast is mother – as is the wooden reel. Indeed – unconsciously – any part *is* the whole. Matte Blanco quotes a schizophrenic patient as saying "my arm is my body" and meaning this literally. This is also an intriguing feature of the mathematics of infinite sets; for example, of the natural numbers which, as Galileo observed, contain a part – the apparently smaller set of squares – which can be considered to be as numerous as the whole.

It is tempting to see the various well-known paradoxes of infinite set theory as expressing aspects of unconscious process. Despite such paradoxes, mathematicians can and do eventually come to agree with one another about mathematical properties, even where they cannot agree about the ontological status of mathematical objects. With agreed meanings of the terms employed, the square root of 2 is either rational or it is not – and ways have been derived of finding out which is the case. But, echoing the theological discussion of icons, mathematicians disagree about the nature of the square root. Is it a human construction or something already existing elsewhere?

I do not myself think it is a very important issue to consider whether mathematics (or art) is a matter of discovery or creation. The philosopher Michael Dummett (1964, p. 509) commented that this particular either/or is a false dichotomy which surreptitiously dominates our thinking. But I am intrigued by the certainty with which others can answer the question and yet do so in contradictory ways. Where does all this certainty come from? According to the analyst Adam Phillips, the psychoanalytic question shifts from asking whether what you say is true to enquiring about what it was in your personal history that disposes you to believe in a particular answer.

Is it that there is some predisposition for some people to see as 'real objects' what others see as ideas and conceptions? In discussing religious experience, William James (1902) referred to a human ontological imagination which he suggested he himself may have lacked. In this, he was impressively unpatronising about other people's experience. This is unlike most of us: although we can sometimes agree about external reality, we do not always find it so easy when it comes to the internal realities others appear sometimes to have created.

The argument seems to me to be of this order. So it may be relevant to consider another context in which notions of internal and external reality are discussed – namely, the work of the so-called object-relations school of psychoanalysis. For instance, Donald Winnicott pointed to an ambiguity in a baby's experience of being breast-fed. The baby begins to believe in an external reality which is encountered as if by magic. Contact of the nipple with the baby's mouth gives the baby ideas. In a sense, the baby creates the object, but it was there waiting to be created.

This is interestingly echoed, in a different context, by Dummett's comment on the perennial debate about the nature of mathematical objects, which he sees as springing into being in response to our own probing. For Winnicott, some babies gain the illusion of finding what was preconceived. But for less fortunate babies, there may be some distress at the idea of there being no direct contact with an external reality. Is the breast there when I want it? Sooner or later ... or, for some unfortunates, never.

One, Two, Three, ...

To go back to the beginning – when there were no words. It is difficult to re-enter the experience of what it was like to be in a pre-verbal state. We not only now produce words, but are in a sense also produced by them. How we make the transition into the world of language can be an important determining factor in our mental health – certainly also in our power to symbolise, to handle the as-yet unknown, to be mathematical.

In the beginning everything is one. Mother-and-babe form a whole, a stable unit. Somehow, sometime, mother becomes other and then there is two – though still as a sort of unit, a pair. But the other may not always be there: the baby cannot then be one with the mother and has to cope with a sense of lack, or – in the metaphoric language of psychoanalysis – with the absence of breast. As Freud (1917/1955, p. 249) memorably wrote, it might then be that "the shadow of the object fell upon the ego". [16]

The lost object may be retrieved in fantasy, in magical wish-fulfillment or in symbolic enactment: *fort–da*! The mother-and-child was a favourite iconic theme which was to be taken up by most Renaissance painters, as well as by René Magritte in a strikingly-titled painting *The Mathematical Mind* (*L'Esprit de géométrie*). This latter reverses the usual image to show a man cradling his mother – thereby making some unconscious overtones in the original theme disturbingly explicit (Figures 6 and 7).

Figure 6: Sandro Botticelli, Madonna del Libro, *1480, Museo Poldi Pezzoli, Milan*

Figure 7: René Magritte, L'Esprit de géométrie, *1937, ©Tate, London 2004*
© René Magritte Estate/ADAGP (Paris)/SODRAC (Montreal) 2005

To recognise difference is to create a boundary, to distinguish inside and out-side, me and not-me. The primitive experience of 'two' is of a polar duality. When does two become three? When does a sense of unit and pair move into being able to count on to three and beyond? Well, there is often a father or father-figure in the background, who in some sense eases the baby away from mother. A famous painting of the Holy Family by Michelangelo can be taken to illustrate this: it has Joseph lifting the baby Jesus over the head of Mary. The babe is no longer the object of exclusive maternal attention; the father will play a socialising role (Figure 8).

Figure 8: Michelangelo, The Holy Family, *1456, Uffizi Gallery, Florence*

For the psychoanalyst Jacques Lacan, the infant encounters language through a pre-established symbolic agency which he called the 'Name of the Father'. The intrusive third – whether real or imagined father – is associated by Lacan with this transition to what he called the Symbolic Order. And every word, every symbol, is a step away from mother. Though, like the baby in Michelangelo's picture, we may still tug at her hair.

Winnicott (1971) linked the use of symbols with the first experience of play and locates this in what he termed a 'potential space':

> From the beginning the baby has maximally intense experiences *in the potential space between the subjective object and the object objectively perceived,* between me-extensions and the not-me. This potential space is at the interplay between there being nothing but me and there being objects and phenomena outside omnipotent control. (p. 100; *italics in original*)

According to Winnicott, the baby invokes 'transitional objects', such as comforters, teddy bears, toys, and so on, to mediate between Mummy and Not-Mummy. Such objects are symbols that unite what are to become two separate things (echoing an original meaning of the Greek *simbolein,* 'to put together'). They are also a step into counting, akin to the Lacanian intrusive third that forces a break in the stable duality. [17] The 'third area' is where we experience our first use of a symbol and our first experience of play. It is such mediation between fantasy and reality that seems to be invoked in the experience of creativity in any field.

Playing and Reality

Mathematical objects may also be seen as transitional objects in Winnicott's sense. As a mathematician, Philip Maher (1994), has observed [18]:

> If we accept the view that one's mathematical reality is an instantiation of one's potential space that occurs when one is doing mathematics then the objects in this psychological space – the mathematical objects one plays with [...] – *function as* transitional objects. From this perspective there is little psychological difference between, say, a teddy bear and a self-adjoint operator [...] (p. 137)

There seem to be different ways in which we cope with the transition between an original inner world and an outer world into which we eventually have to immerse ourselves. The third world, which mediates between these two, is where much of our cultural experience is rooted. Our initial experience of it, it is being suggested, determines to some extent how comfortable we are with symbolic representations of experience, how much we can trust 'abstract' objects to provide what we want from them.

Children play with their transitional objects, whether these are actual toys like dolls or virtual ones like numbers. Any toy involves participation, in the sense that it is often constructed by a child out of ready-made things that might be lying around or thoughts that have come to be noticed. A toy can disconnect the child from the purely functional world, so that in some private imaginative world the child becomes more aware of what she is doing. A toy is something you can play with. And play is the imposition of the imagination on the fabric of the real world.

Play is also seen by Winnicott (1971) as the root of much cultural experience, be that art, science – or mathematics: "Cultural experience begins with creative living first manifested in play" (p. 100). At best, potential space is enriched by the baby's own creative imagination. At worst, the baby is unable to trust his own experience and becomes over-dependent on others. Winnicott quoted another analyst, Alfred Plaut (1966), "the capacity to form images and to use these constructively by re-combination into new patterns is – unlike dreams or fantasies – dependent on the individual's ability to trust" (p. 130). The most difficult students to teach – in art or mathematics classrooms – are those who have lost trust in their own capabilities.

On the seashore of endless worlds, children play.

This quotation from Rabindranath Tagore opens Winnicott's remarkable article (1967) on the location of cultural experience. For the psychoanalyst, the sea and the shore represented "endless intercourse between man and woman" (p. 368). The child of this union comes up out of the sea and lands up on the shore. And holds both sea and shore in imaginative play.

This places the origin of all subsequent 'playful' cultural activity in this first experience. With the suggestion that there is not much inner difference between various adult cultural activities. For Willem de Kooning, according to Harold Rosenberg (1964, p. 115), the objects he depicted "carry emotional charges of the same order as numbers, mathematical signs, letters of the alphabet".

The Hidden Order of Art

There are various ways in which people have been able to link mathematics with other activities. Mathematicians have joined friends of other interpretable objects. Thus, after the painter Maurits Escher discovered the Alhambra decorations in the 1920s (they had not been previously specially noted in guidebooks), mathematicians took an interest in these Islamic designs. And then they took an interest in the more complicated geometry to be found in Escher's own work.

A different sort of link with mathematics was made by a theorist of art education, Anton Ehrenzweig (1967). In his day, many art teachers were concerned that the transition to puberty involved a so-called representational crisis in which it was supposed that children who had been freely painting suddenly became acutely self-conscious and self-critical about their attempts to represent what they saw in a more photographic way. Ehrenzweig offered art teachers a way of understanding this in psychoanalytic terms.

Although he agreed that unconscious process was wholly undifferentiated, he also held that it was not chaotic. Whereas there are no conceptual distinctions in the unconscious, there is, he suggested, some *perceptual structure* and it was this that constituted, for him, the 'hidden order' of art. This could be invoked by tapping powers of abstract thought, which he held became accessible in the years before puberty. Frustrated by their inability to represent realistically the outer world, children could find satisfaction in abstract representation of their inner world.

'Perceptual structure' condenses experiences from early childhood, experiences which are unconscious in the sense that one is normally unaware of them and perhaps has no adequate language to describe any recaptured or transformed manifestation. One example is the way we do not normally need to be aware of how our body fits into its surrounding space. Is pleasure in the curve of a piece of sculpture, or – more concretely – delight in the feel of a shoulder, a re-awakening of a submerged experience of mother's breast?

Mathematicians are used to working below a mental threshold of consciousness in which things are known but not yet thought. What they sometimes loosely refer to as intuition is – perhaps more usefully – described by Christopher Bollas (1987) as an "unthought known". [19] Ehrenzweig's discussion of this sort of unconscious scanning invoked the work of the few mathematicians who have tried to describe the process of mathematical creation.

But we might distinguish the smooth problem-solving process suggested by the influential mathematician Henri Poincaré from the more emotionally charged, psychoanalytic version of unconscious process. Wilfrid Bion, for instance, invoked Kleinian theory to suggest that creativity (whether in art or mathematics) involves a return to the so-called paranoid–schizoid position – as in the dance of the god Shiva, for whom there is no creation without destruction. [20]

Unconscious perceptual structure may be triggered in various ways. But Ehrenzweig's point was that this structure was the hidden scaffolding, the unarticulated (but not necessarily repressed) experience, that was particularly re-awakened by abstract art. So art teachers were encouraged to use the latency period, on the one hand, in order to work more formally with abstract elements like points, lines and circles and, on the other, to teach drawing techniques in a direct and formal way.

There are obvious corresponding implications for mathematics teachers, but these have not been so well worked out. There has always been a widespread view that arithmetical processes, for instance, are best mastered through practical applications. At one time, this meant working through interminable calculations of artificial shopping bills – and there are still many contemporary equivalents. But there are alternative approaches that strongly confirm Ehrenzweig's different point of view. These also indicate that – in the case of mathematics at any rate – abstract symbols can be confidently manipulated (the hidden metaphor is significant) at an early age.

In general, such symbols are hollow: Ehrenzweig referred to the "full emptiness" of abstract art. This means that they can act as condensations. For example, points, lines and circles – the stuff of geometry, as well as the basic elements of much abstract art – matter for various reasons, but also because points lie on lines and lines pass through points or touch circles, and these may be symbols for what I myself lie on, pass through or touch.

Structure on Structure

The power of mathematics lies in its abstract generality. And it sometimes seems that mathematicians can be quite voracious in their exercise of this power: they will 'mathematise' wherever they can. This tendency was, of course, the key to scientific mastery of the natural world. It also holds out hope that its use might eventually lead to a proper understanding and control of political and economic affairs. But, despite such achievement and promise, it is often, quite justifiably, held in some suspicion. There can be something bloodless, something reductive, about some of its applications.

One of the earliest uses of the verb 'mathematise' is ascribed in the dictionaries to the nineteenth-century diarist Henri-Frédéric Amiel (1885), who criticised a contemporary for "mathematicising" morals. When reviewing Taine's contemporary history of English literature, Amiel echoed the earlier scorn of the Romantic poets at Haydon's dinner party:

> instead of animating and stirring, it parches, corrodes, and saddens [...] It excites no feeling whatever; it is simply a means of information [...] giving us algebra instead of life, the formula instead of the image, the exhalations of the crucible instead of the divine madness of Apollo. Cold vision will replace the joys of thought, and we shall see the death of poetry, flayed and dissected by science. (pp. 181-182)

It seems that anything can be mathematised. A search on the internet reveals about two thousand applications of 'mathematisation' to nature, space, economy, psychology, reckoning, motherhood, nonsense, paintings, symphonies, poems, sea-shells, pottery, agricultural practices, ...

David Wheeler (1979/2001) suggested that mathematisation [21] was most easily detected in situations where something quite different was being turned into something which was immediately recognisable as mathematical. He gave as examples a young child playing with blocks and using them to express awareness of symmetry or an older child experimenting with a geoboard and becoming interested in the relationships among the areas of the triangles he can make.

> We notice that mathematisation has taken place by the signs of organisation, of form, of additional structure, given to a situation.

> I use these tenuous clues to suggest that: mathematisation is the
> act of *putting a structure onto a structure.* (p. 51; *italics in original*)

Wheeler's examples were taken from acceptable educational contexts where it seems that the structured blocks and geoboards are specifically intended to have a mathematical structure laid upon them. In the case of a further example he gave, that of an adult noticing a building and asking himself questions about its design and so on, the issue is perhaps more complicated. The determined teacher may invite students to look at a building, in order to stimulate some mathematical thinking. But it should be recognised that this may be a form of iconoclasm – the mathematical structure being achieved and appreciated at the expense perhaps of whatever other qualities (or iconic significance) the building may have previously possessed.

This is an issue that delicately lurks within the practice of identifying (ethno-)mathematical structures laid on the structures of the various craft works of a particular culture.

Words or Things

How do you say about anything more than it says itself?

The difficulty of interpreting artifacts on their own terms may be illustrated by the case of the amazing stone balls that have been found in various neolithic sites in Scotland. There are about four hundred of these now in various museums; they are 7–10 cm in diameter and are carved so as to form a roughly symmetric shape with a number of knobs. Half of the known examples have six knobs – that is, they are more or less cubes with six curved 'faces'. There are then examples of stones having three, four or five knobs, as well as ones with various further numbers, including as many as eighty. One or two of the stone balls are very intricately engraved with spiral patterns (Figure 9).

Figure 9: The Towie ball, c. 2500BC, © The Trustees of the National Museums of Scotland

Three unfinished balls show that the stone was shaped to a sphere before carving of the knobs began. It is not known what the balls would have been used for; early speculations include suggestions that they were used as missiles, as devices to whistle in the wind when thrown, as balls in some game or as some sort of regal orb.

For Keith Critchlow, an architect who has been interested in recovering 'lost knowledge', the balls were manifestations of a neolithic delight in the objects for their own sake. He found that the stones included examples of all five regular (Platonic) solids, as well as four of the so-called semi-regular solids. According to Critchlow (1979), these examples suggest a coherently worked out geometric awareness of symmetry more than a thousand years before the ancient Greeks developed their enumeration of the regular solids.

Critchlow suggested that the stones objects provide "as clear and con-cise a statement in their own terms as any that could be made in either ver-bal or written form" (p. 149). But he is also inevitably trapped into trying to translate the statement made by the stones in some way. It is ironic that in wishing to counteract the tendency to seek cultural origins in Hellenistic terms, Critchlow interprets the mathematics involved in a strictly Euclidean way. But if the stone balls are to be interpreted mathematically, then they might as well be done so in more general terms. For example, they might be seen to provide interesting solutions to the problem of finding *k-maxi-mal* sets of points on a sphere – that is, roughly speaking, the problem of finding arrangements of a number of points on a sphere that spread the points as far apart as possible.

There are numerous further examples of objects that can now be re-interpreted, if we wish, in various mathematical ways. But, as Tamen (2001) suggests:

> it is conceivable that what most of us would call interpretation could be described by others precisely as a strange, incomprehen-sible exercise, perhaps akin to what we in turn would consider a rain dance, or a complex ceremony in a physics laboratory. (pp. 131-132)

Wrought-up Things

These neolithic objects preserved in museums, and analysed mathematically, have certainly lost any original iconic function. What about our own current iconic objects? These will nowadays be different. But our special objects, our *anathemata,* can speak to us in ways which, if not the same as those that Theodore tried to describe, are at least similar. If God is not now always in the eye of the beholder, there is often some mystery that is. And we may interact with it unconsciously, rather like the way we can sometimes with people. Psychoanalysts have technical words with which to analyse this sort of communication, one that is ordinarily referred to in terms of empathy or intuition. But the important issue is that almost everyone experiences it in some way or other, *though they may not always know they do.*

To know (some of) what it is they know is the lot of mathematicians who need to be continually aware of awareness. This trained, self-reflective activity may be pleasurable and life-enhancing or it may be painful and life-denying. Or it may oscillate between these two. There is no need to settle for one aspect rather than the other, any more than there is to settle the argument between iconophils and iconoclasts: or, indeed, between those who would see a mathematical object as a human construct or as a discovered element in some world or other. Away with such distinctions! What is at issue is whether we can let our experience become iconic (in the original sense of that word).

Artists have always tried to capture this for us. For instance, in her novel, *Adam Bede,* George Eliot charted some subtle complexities of human relationship. In a few beautifully written pages, Eliot (1859/1961) set the scene in which Adam will find Hetty in the garden and imagine that his feeling for her is reciprocated. Adam finds her gathering fruit. He is overcome by the sense that they are sharing a mutual, as-yet unspoken love. Unfortunately, he is mistaken: she has been, and is, thinking of another. But still unaware of this, he thinks he detects a sign, "a slight something", that his love is returned. The memories of the first moment of shared love – Eliot called it "the time that a man can least forget in after-life" – are then compared with the vanished memories of childhood in one stirring sentence.

> So much of our early gladness vanishes utterly from our memory: we can never recall the joy with which we laid our heads on our mother's bosom or rode on our father's back in childhood. Doubtless that joy is wrought up into our nature, as the sunlight of long-past mornings is wrought up in the soft mellowness of the apricot, but it is gone for ever from our imagination, and we can only *believe* in the joy of childhood. (p. 215; *italics in original*)

It is, of course, Eliot's achievement that thanks to her we might imaginatively re-experience some of this joy that is "wrought up into our nature". But what do we make of *wrought up?* It seems that the writer suddenly leaves the very physical, tactile images that describe the garden and uses a word that describes a process: one that might be understood by the mathematically sensitive in different (perhaps topological) terms. If so, this might be to mathematise in an enhancing way, one which gives something back to the image of the apricot's wrought-up mellowness.

There may, of course, be images that assert a mystery that cannot be analysed so easily: for example, the strange, self-conscious photograph of a broom in an open doorway, by the early pioneer of photography, William Henry Fox Talbot. [22]

Vibrating Strings

The artist Max Bill (1949/1993) ignored domestic objects like brooms when he listed such other things as the mystery enveloping all mathematical problems, the inexplicability of space, the remoteness or nearness of infinity, the disjunctive and disparate multiplicities constituting coherent and unified entities, and so on. These metaphysical brooms:

> can yet be fraught with the greatest moment. For though these evocations might seem only the phantasmagorical figments of the artist's inward vision they are, notwithstanding, the projections of latent forces; forces that may be active or inert, in part revealed, inchoate or still unfathomed, which underlie each man-made system and every law of nature it is within our power to discern. (p. 8)

How one sensitises to such forces is, of course, part of a psychotherapist's training. But artists, and indeed mathematicians, often train themselves. People may attend in their chosen disciplines to different objects, but the way they attend is often very similar. An early, remarkable account was given by Denis Diderot (1769/1966). In a fictional dialogue, he talks with his contemporary, Jean D'Alembert. The latter suggests that if an object is to make any sense, it has to remain under scrutiny while the intellect affirms or denies certain qualities of the object. Diderot, in character, agrees and goes on to describe the process.

> That's what I think, and it has sometimes led me to compare the fibres of our organs with sensitive vibrating strings. A sensitive vibrating string goes on vibrating and sounding a note long after it has been plucked. It is this oscillation, a kind of necessary resonance, which keeps the object present while the understanding is free to consider whichever of the object's qualities it wishes. But vibrating strings have yet another property, that of making others vibrate, and it is in this way that one idea calls up a second, and the two together a third, and all three a fourth, and so on [...] (p. 156)

In reality, D'Alembert could mathematise vibrating strings, while Diderot was able to use this to give an excellent metaphorical account of a process we might now describe in other terms. If the phenomenon described can be observed between resonant strings, then why, asked Diderot, cannot it take place between "living and connected points, continuous and sensitive fibres?" (p. 156).

Between people, or between a person and an object? A sensible object? An artifact or a mathematical object? The central issue that I have been trying to hint at, rather than assert too explicitly, is that in each case it is whether the vibrations reach the mystery and whether we can participate in, and perhaps venerate, this mystery.

Psychoanalysts have ways of describing such a process. For instance, we may project something of our own into an object – a person, some inanimate thing, a mathematical diagram – and this may then speak back to us in an accessible (positive or negative) way. In considering Marcel Duchamp's exhibition in the 1920s of a porcelain urinal (an early example of 'conceptual' art), Adrian Stokes (1965, p. 13) suggested that, "We, the spectators, do all the art-work in such a case, except for the isolating of the object by the artist for our attention". As always, Blake (1810/1967) put it in his own stirring way:

> If the Spectator could enter into these Images in his Imagination, approaching them on the Fiery Chariot of his Contemplative Thought, if he could Enter into Noah's Rainbow or into his bosom or could make a Friend & Companion of one of these Images of wonder [...] then would he arise from his Grave, then would he meet the Lord in the Air & then he would be happy. (p. 162)

As an exercise for the reader, here are two mathematical "Images of wonder" – one probably quite familiar, the other perhaps not. [23]

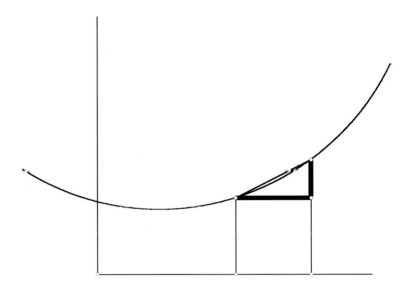

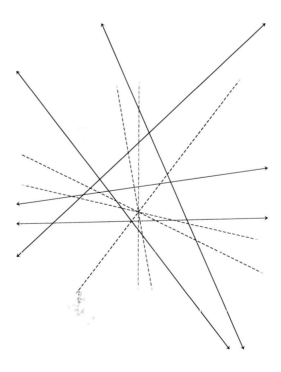

The Mystery of Things

When Shakespeare's Lear finally begins to understand what has happened to him, and what he has caused to happen to others, he clutches briefly at the hope of some idyllic, though imprisoned, last years with his re-discovered daughter. He envisages the two of them singing like birds in a cage. They will, he says, "pray, and sing, and tell old tales, and laugh at gilded butterflies [...] *And take upon's the mystery of things as if we were God's spies"*.

"What things?", asks Christopher Bollas (1999, p. 195). [24]

> The things that live as effects, in the subjects who cultivate them, in the objects presumed to contain them, in the receivers assumed to know them not for what they are, but the familiar movement of the 'are not'. Not the themes of life, the plots of the novel, the urgent reports of the analysand, but the forms of life.

"What mystery?", he goes on to ask.

> An unanswerable, perhaps presiding question. What is the intelligence that moves through the mind to create its objects, to shape its inscapes, to word itself, to gather moods, to effect the other's arriving ideas, to ... to ... to?

Figure 10: Paul Cézanne, Still life with apples and oranges, *c. 1899, Musée d'Orsay, Paris*

Notes

[1] St Saviour in Chora (i.e. in the country, so outside the original city) dates from the early fifth century. The mosaic of Christ Pantocrator (the Almighty) in the southern dome of the inner narthex was part of an extensive fourteenth-century restoration. The frescoes and mosaics were – in an inevitable iconoclastic act – whitewashed over when the church was converted into a mosque. It was handed back in the 1950s to the Byzantine Institute of America and has been brilliantly restored.

[2] St Theodosia, an ardent iconophil nun, was stabbed in the throat with a ram's horn by the irate soldiers. She is portrayed in a fourteenth-century icon (now in the British Museum), holding what is also supposed to be a small copy of the Chalke mosaic.

[3] The first Church Council of Nicea in 325 declared that the Son was begotten of the 'same substance' (*homo-ousion*) as the Father. Those who wished to emphasise Christ's human nature assumed he was of 'like substance' (*homoi-ousion*). Sceptics have made great play with the 'missing iota'.

[4] St Theodore was born in 759 (in the reign of Constantin V, the second icono-clast emperor). Theodore was involved in controversies all his life and was exiled three times – the last for defending the veneration of images.

[5] In preparing this chapter, I have been much indebted to Tamen's stimulating and highly original book.

[6] There was plenty to drink, and much merriment, among these 'friends of poetical objects'. For a recent account of the famous dinner, see Hughes-Hallet (2000).

[7] An original sense of the word 'icon' is being invoked here, not that of current semiotic usage where it might be a photograph or painting rather than the rainbow itself that would be the icon. It was the divine presence that was venerated, not the mosaic material as such, even if it was held that the material contained – or *was* – the presence.

[8] It seems ironic that Cantor always denied the reverse, namely that you might see the One as a Many. Though he defended transfinite numbers against much scepticism, he was aggressively critical of the notion of infinitesimals, which were, in a sense, a mathematisation of Blake's "world in a grain of sand".

[9] In the preface to his poem, Jones distinguished between *prudentia* and *ars*. The first involves our intentions and dispositions and our final condition. But artifacts are, he claimed, already complete: they clamour for attention, *now*.

[10] Balint had earlier made an interesting distinction between those he called *ocnophils*, who are attached to objects and are not at ease with open spaces, and *philobats*, who are at an opposite extreme. See also Chapter 7 of Klein (1987).

[11] An echo of Plato's view of mathematics occurs in a 1980 statement of aims for primary school mathematics in Saudi Arabia. This was compared with a contempo-rary statement for the UK by Geoffrey Howson (1984). He summarised the Saudi aim as: "to move children's thoughts from the concrete world around them to the abstract, [in a movement] from matters temporal to thoughts of things spiritual" (p. 41). St Theodore, at any rate, would have approved of this point of view:

"Generalities are seen with the mind and thought; particular individuals are seen with the eyes" (1981, p. 83).

[12] Blake's comment on seeing *through* was that you are led "to believe a lie when you see with, not thro', the eye" (*The Everlasting Gospel,* lines 103-104). G. Spencer-Brown has noted that *theatre* and *theorem* have the same root, both suggesting display or spectacle – in the mathematical case, being 'seen' with in-sight (see Keys, 1971, p. 35).

[13] The following diagram illustrates the proposed way of 'seeing' the staircase more directly. But it was not included in the main text, so that the reader 'gazing' at the first diagram would not be distracted. This example of a pictorial proof is taken from Brown (1999, p. 35).

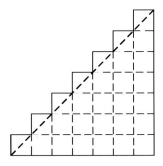

[14] Compare this with Tahta (1994). For an account of the way subjective interpretations might be validated by 'resonance' in a 'community of practice', see Mason (2002).

[15] See also an application of Matte Blanco's ideas in Mordant (1993).

[16] Freud suggested that after the loss of a loved object, for whatever reason, 'free libido' is then displaced onto another object or has to be withdrawn into the ego. In the latter case, the ego may then become identified with the abandoned object – hence the shadow and melancholia.

[17] Pimm (1994, pp. 119-120) offers an anecdote where a young child counts 'one, two, two' that seems to shed some light on this.

[18] In this piece, Maher neatly links the views of Winnicott and Lacan.

[19] According to Bollas (1987), "We need a term to stand for that which is known but has not yet been thought […] There is in each of us a fundamental split between what we think we know and what we know but may never be able to think." (pp. 280, 282). The notion of an 'unthought known' is also invoked in his discussion of the aesthetic moment, which "holds self and other in symmetry and solitude [with] a deep rapport between subject and object [and] the generative illusion of fitting with an object" (p. 32).

[20] Melanie Klein saw what she termed the 'paranoid–schizoid position' as a state which is specific to the first four months of life, but which may often be reverted to in later life. In this position, there tends to be a split between 'good' and 'bad'

objects. This is seen as a chaotic state which needs to be resolved by a transition to the so-called 'depressive position', in which good and bad are seen as one and in which we can accept love and hate of the same object. Bion introduced the idea that in creative problem solving one might have to revert temporarily to the previous state. See Skelton (1993).

[21] He published a number of articles on the theme of mathematisation: see, for example, Wheeler (1982).

[22] Fox Talbot was an early pioneer of photography. He was also an amateur mathematician and regularly contributed to mathematical journals throughout his life; he was elected a Fellow of the Royal Society for his work on elliptic integrals. He invented a photographic process in the 1830s which involved printing on paper, rather than Daguerre's earlier, but more cumbrous, production on a silver plate. The photograph of the broom, entitled *The Open Door,* appeared as Plate 12 in his book, *The Pencil of Nature,* published in 1842. Early photographs were shadowy and somewhat mysterious. Later precision and the elevation of photography into art have been seen by Marxist critics as a form of iconoclasm: see Chapter 6 of Mitchell (1986).

[23] The first diagram is part of the classical geometrical introduction of the derivative of a function as the slope of a tangent. The second may require some elucidation: it illustrates a remarkable theorem about five special lines associated with the five complete quadrilaterals obtained by leaving out one of five lines in turn. For each quadrilateral, there are three diagonals; it was shown by Newton that the mid-points of these diagonals lie on a line, which has been called the *diameter* of the quadrilateral. The diagram illustrates the theorem that asserts the five such diameters all meet in a point. This was at one time ascribed to Fox Talbot (see note [22]), but was in fact first published by a French mathematician, Olry Terquem, in 1845. The proof is, as they say, left to the reader. There are some interesting generalisations of this wondrous result.

[24] The last chapter of Bollas's book is especially recommended for its psychoanalytic version of Diderot's vibrating strings. See also the clear account given in Bollas (2002). Also, note the Shakespearean echo once again in Virginia Woolf's (1931/1950) novel *The Waves.*

> 'Like' and 'like' and 'like' – but what is the thing that lies beneath the semblance of the thing? [...] There is a square; there is an oblong. The players take the square and place it upon the oblong. They place it very accurately; they make a perfect dwelling-place. Very little is left outside. The structure is now visible, what is inchoate is here stated; we are not so various or so mean; we have made oblongs and stood them upon squares. This is our triumph; this is our consolation. (p. 116, *Rhoda speaking*)

> So now, taking upon me the mystery of things, I could go like a spy without leaving this place, without stirring from my chair. I can visit the remote verges of the desert lands where the savage sits by the camp-fire. Day rises; the girl lifts the watery fire-hearted jewels to her brow; the sun levels his beams straight at the sleeping house; the waves deepen their bars; they fling themselves on shore; back blows the spray; sweeping their waters they surround

the boat and the sea-holly. The birds sing in chorus; deep tunnels run between the stalks of flowers; the house is whitened; the sleeper stretches; gradually all is astir. Light floods the room and drives shadow beyond shadow to where they hang in folds inscrutable. What does the central shadow hold? Something? Nothing? I do not know. (p. 207, *Bernard speaking*)

Chapter ω

Aesthetics and the 'Mathematical Mind'

David Pimm and Nathalie Sinclair

> Listen, reader – the clock is ticking. Have you shored up your
> days against pleasure and the strange? (Mark Cochrane, in Cran,
> 2002, p. 82)

Many authors have explicitly sought to evoke or identify the 'universal' in
discussing aesthetic aspects of mathematics, a quality that is regularly
claimed for mathematics itself. Such discussions at times allude to the important values of distance and detachment on the one hand and to the qualities of certainty and perfection on the other. We also believe these to be
worthy themes of attention, as this chapter bears out. Our approach, however, invokes psychological themes, ones that lead to, among other things,
seeking the natures of and ascribing origins to those impulses that make
humans both long for and seek out detachment and certainty – as well as
exploring why some choose to place them so centrally and unequivocally
within mathematics.

Such a distanced, 'objective' view of aesthetics involves a way of talking
and writing about mathematical experience that mathematician Gian-Carlo
Rota (1997) has trenchantly termed a "copout". In a chapter called 'The phenomenology of mathematical beauty', Rota writes:

> Mathematical beauty is the expression mathematicians have invented
> in order to obliquely admit the phenomenon of enlightenment while
> avoiding acknowledgement of the fuzziness of this phenomenon.
> [...] This copout is one step in a cherished activity of mathemati-
> cians, that of building a perfect world immune to the messiness of
> the ordinary world, a world where what we think *should* be true
> turns out to *be* true, a world that is free from the disappointments,
> the ambiguities, the failures of that other world in which we live.
> (pp. 132-133; *italics in original*)

Rota suggests that mathematicians acknowledge a theorem's beauty when
they see how it 'fits' in its place, how it sheds light around itself. Within his
sense of the mathematical aesthetic, the importance of personal understanding is elevated. A proof is beautiful not because it is pure or detached from
human emotions and needs, but because it gives away its secret or leads the
mathematician to perceive the apparent inevitability of the statement being
proved after the event. Rota proposes that it is precisely this phenomenon
of individual enlightenment that keeps the mathematical enterprise alive.

There is a significant danger in construing aesthetics purely as a socio-cultural construction (let alone an 'objective' one), not least from the attendant risk of losing sight of the richness and vividness of experience that individual mathematicians sometimes report (e.g. Thurston, 1994, or Wiles, quoted extensively in Singh, 1998). Here is Fields medallist Alain Connes in conversation with neurobiologist Jean-Pierre Changeux (1995) about Hadamard's four stages of invention and the role of emotions in mathematics in relation to emergence from the *incubation* period:

> *Changeux:* A kind of "pleasure alarm" goes off, in other words, rather than a danger alarm, signaling–
> *Connes:* That what's been found works, is coherent and, one might even say, aesthetically pleasing. I'm certain that this pleasure is analogous to that experienced by painters the moment they find a solution, the moment they see that a canvas is perfectly coherent and harmonious. The mathematical brain must function in the same way. (p. 81)

Slightly earlier in their dialogue, Connes had observed about Hadamard's fourth stage (*verification*):

> *Connes:* At the end of the incubation period, if one is lucky, one experiences the illumination he describes. The final phase, verification, begins once illumination has taken place. The process of verification can be very painful: one's terribly *afraid* of being wrong. Of the four phases it involves the most anxiety, for one never knows if one's intuition is right [...] But the moment illumination occurs, it engages the emotions in such a way that it's impossible to remain passive or indifferent. On those rare occasions when I've actually experienced it, I couldn't keep tears from coming to my eyes. (p. 76; *italics in original*)

Contemporary views of aesthetics often attempt to combine elements of pleasure with aspects of sensory perception. We are well aware that the notion of aesthetics by itself is far from synonymous with that of 'pleasure', especially perhaps when speaking of mathematics. But it nonetheless seems to us worth asking what are some potential sources of pleasure for mathematicians (or by enquiring, to invoke that more charged term, what are their objects of desire). Pleasure, longing or desire, however, as we know from more than a century of psychoanalytic concern, are never straightforward. And all three involve unconscious thought.

One key element in some of the chapters in this book has been attention to the role of unconscious processes in mathematical life and thought, as opposed to solely conscious, exclusively cognitive, 'detached' and more 'objective' accounts. Chapter 9 made mention of Philip Maher's (1994) linking of teddy bears and self-adjoint operators: transitional objects both, merely functioning at different levels of play and psychological development. And the opening of Chapter 4 described the young François Le Lionnais engaged with his number 'toys' at the family table. Possession – '*my*' hemiolic crystal', *'ma courbe'*; in other words, the having and holding of both first and sub-sequent possessions – is of central importance. When reading accounts of the mathematician's proclaimed detachment, it is worth remembering that individuals are never detached about their transitional objects. [1]

In *The Ruler, the Compass and the Couch: Mathematical Pleasures and Passions*, mathematician Nicolas Bouleau (2002) has written an engaging and insightful set of psychoanalytically inspired essays on mathematics and mathematicians. In the first of these, entitled 'l'inconscient mathématicien' [2], he details the considerable amount of writing by mathematicians in the nineteenth century involved with the notion of unconscious processes and their role in relation to creativity in particular. [3]

In the opening preamble to his book, Bouleau observes:

> Poincaré, in the famous recounting of the circumstances of his dis-covery of Fuchsian functions, clearly argues for an active role for the unconscious. It should be noted that, a century before, in a foreshadowing text, Laplace offered a similar notion in the lan-guage of the period. But the interplay between the imaginative freedom of the unconscious and conscious logical control is described with astonishing care by Poincaré, in terms close to those of Freud, were it not for the absence of the notion of pleasure, which was present in Laplace, for which aesthetic taste was the substitute. (pp. 10-11)

Bouleau's comment, echoing Rota's in terms of alluding to the smoothing and even deflecting function of much talk about aesthetics in mathematics, can serve as a reminder about the presence of individual investments in whatever accounts are provided. In the past couple of decades, there have been an increasing number of published interviews with mathematicians, whether undertaken by other mathematicians or by outsiders to the profes-sion. [4] When reading such accounts, it is worth bearing in mind the vested nature of such narratives, while also keeping an understanding ear out for more subterranean and second-order readings and potential themes.

The need for an under-standing ear as well as an under-listening one is raised in a chapter by Valerie Walkerdine (1988) called 'Pleasure and the mas-tery of reason'. While describing the entry of the 'idea of beauty' into claims for mathematics education (for instance, "children will derive pleasure from the purity and order which they discover", p. 189), she goes further, identi-fying mathematical discourse as the product of desire and spelling out some of the costs (such as anxieties and fears) involved in its mastery. In so doing, and by making use of the work of psychoanalyst Jacques Lacan, she draws attention to "that universal mathematization which Lacan himself designates a fraudulence of certainty" (p. 192). Walkerdine writes strikingly about how the underlying 'Other' of mathematics is *un*certainty, *dis*order, *ir*rationality, being *out of* control. She observes that the symbolic system that mathematics exemplifies "is not constituted out of certainty, but produces certainty out of a terror" (pp. 199-200).

In the above, it seems apparent that two differing accounts of 'unconscious' processes are at work. Poincaré's (and subsequently Hadamard's) appeal to creative unconscious processes in mathematics is evidently not the same as the darker and more chaotic unconscious alluded to by Walkerdine (relying on Lacan) nearly a century later. Poincaré's observations might now be described by some as involving 'pre-conscious' (to use Freud's later term) rather than 'unconscious' (involving repression) mental activity. Others might draw attention to the soothing nature of Poincaré's account, according to which the most unsettling characteristic of what he terms 'the subliminal self' is simply its freedom – the absence of any discipline and the disorder of chance. On the other hand, Walkerdine's allusion to 'terror' in connection with mathematics might be read as over- or even melo-dramatic. [5] But what is clear is that in any account involving pleasure, the nature of that pleasure will depend in part on the specific unconscious involved.

So what sorts and which forms of pleasure does mathematics afford? Alternatively put, what is mathematics's debt to pleasure? Specifically, what is the pleasure of the mathematical text? Might there even be an aesthetics of mathematical pleasure to be developed? Without any further introduction, there are three sites we propose to explore further in this chapter. The first two, discussed in the next section, are linked (desire for distance and detachment, longing for certainty and perfection) and are probably familiar and expected themes. The third (solace in melancholy) is likely quite the opposite, but it takes us into significant historical discussion about the nature of the 'mathematical mind'. We shall also endeavour to bear the duals, the opposing 'Others', of this trio in mind as well: detachment *and* passion [6], certainty *and* tentativeness, melancholy *and* joy, as well as those more established pairs of discovery *and* creation, abstract *and* concrete, general *and* particular, control *and* its loss.

Detachment and Certainty

> Perhaps there is a perfect detachment.
> God knows, I want to believe in things.
> (Patrick Lane, 1982, p. 44)

Detachment, disinterest and 'aesthetic distance' arise over and over as important concepts – although ones difficult to define – in broad discussions about aesthetic experience and aesthetic judgements. These concepts have helped philosophers of art attempt explanations of how people can have aesthetic experiences prompted by, for instance, a dangerous and unpleasant fog at sea or a violent, ugly depiction of war. For Kant (1790/1987), disinterest – that is, separation from personal beliefs, passions and commitments – was essential to aesthetic judgement. In his *Critique of Judgement,* he attempted to provide a way of distinguishing judgements of taste from other types of judgements. This was done by construing them as those made without interest in the real existence of the object perceived, and hence without interest in the relation of the object to one's own goals and ends. [7]

While a number of philosophers (see Dewey, 1934, and Polanyi, 1958, in particular) have argued that Kantian disinterested judgements are impossible for humans, others continue to argue that aesthetic judgements require some kind of distance, as seen in Bullough's (1963) *psychical distance* or Beardsley's (1982) *detached affect*. [8] Furthermore, the non-utilitarian quality of art works (which was an early but sustained misinterpretation of Kant's notion of disinterest) continues to be a focus of discussion in the art world.

This supposed non-utilitarian quality of art has been the basis on which some mathematicians have compared mathematics and the arts. Recall from Chapter α, for instance, Hardy's insistence on the 'purity' of his work, on the fact that his discoveries had no practical application, though Jonathan Borwein declared in Chapter 1 that he thought this had been overstated. Borwein and Jackiw in their respective chapters in this book both explicitly mentioned the detached or disinterested aspect of human mathematical experience.

In discussing the relative value and importance of making explicit a mechanism already implicit (albeit unconscious) in the mind, mathematician René Thom (1973) enquired:

> How can the thinker in some way detach himself from his own thought, visualise it abstractly, independent even of the content of the thought? (p. 199)

Thom accepted its necessity to some extent, while drawing attention to its opposite, to attachment as a significant mathematical process as well:

> Certainly, this detachment is a necessary step in the process of mathematical reasoning: but the inverse operation, which is the reabsorption of the explicit into the implicit, is no less important, no less necessary. (p. 199)

The very process of mathematisation can be specified in terms of its plurality of detachments; it is the detachment from specifics through generalisation, the detachment from real-world referents and details that would serve to define a particular situation, as well as the detachment from personal connections. Furthermore, the codifying and communication of mathematisations, which are often carried out through the writing of proofs, can themselves be seen as exercises in detachment. A proof distances itself from the situations and specific examples to which it applies, as well from the personal commitments and attractions that went into forming it. Chapter 8 also contained reflections on the complex relationship of mathematics to time and human agency.

In fact, Nicolas Balacheff (1988) has written that the very language of conceptual proof demands that the speaker "distance herself from the action and the processes of solution of the problem" (p. 217). In particular, he claims this requires:

- a *decontextualisation*, giving up the actual object for the class of objects, independent of their particular circumstances;

- a *depersonalisation*, detaching the action from the one who acted and of whom it must be independent;

- a *detemporalisation*, disengaging the operations from their actual time and duration: this process is fundamental to the passage from the world of actions to that of relations and operations.
(pp. 217-218; *italics in original*)

One might wonder about the consequences and human costs of this triple detachment. These three 'de-s' imply a reversed norm, namely one where such attachments have all and already been successfully made – attachment to object, to person and to time – something which is not always the case, for example, in individuals with autism. [9] And touch is the human sense that involves the closest proximity and engagement, one that belies or tries to negate distance and detachment. For, as Jonathan Swift (1745/1995, p. 25) remarked, in his booklet *Directions to Servants,* "Feeling hath no Fellow". [10]

As an instance of a possible consequence, what happens when language is used as a tool for logical deductions rather than as a means of communication? What happens when the mathematician writes so as "to conceal any sign that the author or the intended reader is a human being" (Davis and Hersh, 1980, p. 36). Rota (1997) asserts that in this case mathematical proof becomes a form of "pretending", since the language of proof produces a striking gap between "the written version of a mathematical result and the discourse that is required in order to understand the same result" (p. 142). And, thus, understanding is compromised, or rather, exchanged for a kind of functional language, one where: "Clarity has been sacrificed to such shibboleths as consistency of notation, brevity of argument and the contrived linearity of inferential reasoning" (p. 142).

The process of formalisation also requires dehumanisation; the facts must, so to speak, be made to "speak for themselves". (Chapter 8 addressed ways in which conventional mathematical rhetoric helps to create the possibility of mathematical objects being seen as bearers of their own agency.) Rota's notion of proof as detachment from human understanding, or at least involving concealment from it, is related to the values held and promulgated in the mathematics community. After all, one could imagine a world of research mathematics where the journals were filled with proofs that illuminate rather than conceal or proofs that open up and explain rather than codify and hide.

Thus, the world of research mathematics is animated by a specific set of values that distinguish it from other fields of intellectual endeavour. For instance, in Chapter 7, Nicholas Jackiw discusses an ego-driven value, one that the child-like pronouncement "*I* made that" nicely captures. [11] Further examples were given in Chapter 4, where mathematicians strongly identified with the objects of their attention. Additionally, reference to the work and beliefs of the Bourbaki in Chapter 8 draws attention to a certain set of ideological values that exist in mathematics: deductive reasoning is the only true way of achieving satisfactory or trustworthy conclusions and an axiomatic presentation is the only fitting way of offering mathematics.

These examples, as well as others to be found in the previous chapters in this book, highlight the fundamentally aesthetic nature of many values that characterise mathematics and which help determine the very shape of mathematical knowledge and practices. The value of mathematical rationalism shuns – and seeks to detach from – other forms of explanations, forms that betray the presence of their human creators. Even outside mathematics, rationalising can mean seeking logical connections between ideas, thus overcoming the inconsistencies, disagreements or incongruities that may arise from personal interpretations of various situations or ideas. (This echoes Walkerdine's remarks about the 'Other' of mathematics mentioned in the opening of this chapter.) Yet, as Yves Chevallard (1990) has wryly observed, "Mathematics is a perfect example on which a *celebration of ambiguity* could be founded" (p. 8; *italics in original*).

Explaining, ironically enough, can also mean "explaining away": it is possible to rationalise certain decisions or actions by calling on some external, logically attractive principle, one that frequently ignores the principles that may have tacitly but nevertheless actually guided actions. [12] Alan Bishop (1991) writes that we are both guided by and uphold the values of rationalism:

> when we disprove a hypothesis, when we find a counter-example, when we pursue a line of reasoning to a 'logical conclusion' and find it is a contradiction to something known to be true, and when we reconcile an argument. (p. 63)

The aesthetic dimension of rationalism relates to the valuing of the completeness and wholeness that belong to a logical argument. Fuzziness, imprecision and loose ends are banished and are replaced, more and more, with Rota's 'pretend' proofs, which are optimised for consistency, containment and cohesion. The mathematician's desire for such aesthetic qualities may stem from a discomfort with graded truths: logical conclusions, like cohesion and consistency, it is claimed, do not admit degrees. Arguments are, on the one hand, either logical or not, and, on the other, either consistent or not: neither leave room nor exhibit desire for middle ground.

The process of detachment may also enable certain forms of mathematical reasoning. Hendrik Lenstra, in interview with Nathalie Sinclair, explained the role of detachment in the work of algebraists:

> They somehow try to get a handle on their properties by taking a more distant attitude. That is why they introduce more abstract-sounding notions such as rings, ideals. They nevertheless end up being able to prove very concrete theorems about identities that exist or may not exist without really ever exhibiting them, just because they have this superior mechanism.

Lenstra contrasted his very abstract way of working with a colleague's, who has "combinatorial points of view" and thus a more "down-to-earth" way of working. He also hinted at the role that the computer plays in dictating these approaches when he explained why he could work with ideals but his colleague needed to work with polynomials:

> [He needs] to put them into the computer. It doesn't really like to compute with ideals. A computer needs to know what it is doing.

In Chapter 1, Borwein contended that the computer is changing the style of mathematics: the computer forces a more concrete and less abstract or detached way of representing and communicating. Indeed, doing mathematics with the computer has already challenged the complex set of intertwined values held by the mathematics community. (Consider, as examples, Haken and Appel's proof of the four-colour theorem, now some thirty years old, and the more recent claiming of Kepler's conjecture by Thomas Hales. Hales's purported proof, presented for publication in 1998, was originally to have been divided in two between *Annals of Mathematics* and *Discrete and Computational Geometry,* the former to have been published with a disclaimer. For an update on this, see Morgan, 2005.)

The way in which computer-based proofs have sparked new discussions on the nature of truth in mathematics, as well as the interplay between understanding and proving, reflects the way in which issues of detachment and distance may be related to a larger set of mathematical values. As we discuss in the next sub-section, there may be a sense in which desire for detachment and distance gives rise to – and perhaps also results from – a related longing for certainty and perfection.

Certainty and perfection

In this sub-section, we can only start to explore some implications of the quest for certainty [13] in mathematics, as well as the related aesthetic values of 'perfection' and 'order' that mathematics indulges. But, as before with distance and detachment, there is also the question of the costs, obvious or otherwise, of accepting such a grail. [14] The desire for certainty in mathematics has involved sharp and prolonged moves toward abstraction and away from contingent events of the everyday world, while perfection too takes us away from the here and now of the actual into considerations of and comparisons among the possible.

Even if the most devoted Platonist were some day finally to concede the uncertainty of mathematics, there would likely be no escaping the *feeling* of certainty one gets when doing mathematics. [15] There is also the very human *longing* for certainty that sometimes is expressed. Towards the end of his life, Bertrand Russell (1956) wrote:

> I wanted certainty in the kind of way in which people want religious faith. I thought that certainty is more likely to be found in mathematics than elsewhere. But [...] after some twenty years of

very arduous toil, I came to the conclusion that there was nothing
more that *I* could do in the way of making mathematical knowledge indubitable. (p. 53; *italics in original*)

According to mathematician Brian Rotman [16], the desire embedded in the
discourse of mathematicians involves "reason's dream", a powerful fantasy
of permanence and certainty:

The desire's object is a pure, timeless unchanging discourse, where
assertions proved stay proved forever (and must somehow always
have been true), where all the questions are determinate, and all
the answers totally certain. (in Walkerdine, 1988, pp. 187-188).

Walkerdine goes on to assert that, "the result of [this] fantasy is lived as a
fact" (p. 188). Irrespective of the above, mathematics is certainly one significant place where human beings invest their desire for certainty. [17]

Here, we want to draw on Wilhelm Worringer's particular use of the
contrasting terms 'abstraction' and 'empathy', coined at the beginning of the
last century and described in Richard Padovan's (1999) book *Proportion*.
'Abstraction' signals a tendency:

to regard nature as elusive and perhaps ultimately unfathomable,
and science and art as abstractions, artificial constructions that we
hold up against nature in order in some sense to grasp it and command it. (p. 12)

Worringer's contrasting sense of 'empathy' marks:

the tendency to hold that, being ourselves part of nature, we have
a natural affinity with it and an innate ability to know and understand it. Le Corbusier calls this affinity 'an indefinable trace of the
Absolute which lies in the depth of our being'. (p. 12)

The abstraction viewpoint would see mathematics as a human-made creation, a purely artificial construction, a system of conventional signs and
the rules for manipulating them. Nature is thus something we can interpret
though mathematics only because mathematics is a human creation. There
is no need for the fallible human senses and no reference must be made
to natural forms. An empathy viewpoint – also, a Pythagorean one – sees
mathematics as being inherent in nature and distilled out of it by human
reason, through human senses. The shifting appearances of things can be
penetrated by mathematics, which is able to reveal the essential nature of
imperfectly manifested phenomena.

Padovan sums up the paradox encapsulated by these two viewpoints [18]:

No knowledge is possible, unless it comes first through the senses,
but such knowledge is at best uncertain. The certainty of mathematics is due precisely to the fact that it is man-made, the uncertainty of nature to the fact that it is not. (p. 11)

One of the means by which mathematics is believed to be made more certain is through abstraction. As Balacheff noted, abstraction requires, among

other things, both a decontextualisation and a depersonalisation, detaching from the object and from the person. Abstraction is powerful in mathematical and scientific work because it allows one to look for common features across local instances. Then, when we apply our abstract models back to the world and see how they fit, we should eventually get a better sense of how things work. But do we? Or are we, in fact, entranced by the apparent certainty that *only* our abstract models are able to offer? One danger is that we might begin living in such a decontextualised world where we forget about the impact of our abstract models on the real and messy lives of human beings, as Rota cautioned.

Some recent writers have expressed a sense of foreboding about mathematics, with its extreme commitment to abstraction. For instance, Ivan Illich (1994) has pointed to this characteristic as being one of the root causes of social malaise in the modern world. In his lecture, he drew on the work of the social theorist Leopold Kohr, whose ideas on social morphology are structured around the central concern of *proportion* in its classical sense.

Kohr claimed that the ancient Greeks had no conception of individual tones nor of absolute measurements; their world-view was based on a certain sensibility captured by the word *tonos,* which refers to the relatedness of things one to another, to the proportions between humans and nature. Because of this fundamental attention to the relations between things, *tonos* demands a concern for appropriateness. In his talk, Illich argued that the growing mathematisation of science and the desire to quantify – to separate, abstract, detach – has reduced our capacity to judge appropriateness and to attend to, as well as to return to, the particularities and the proportions of local meanings.

No one can deny the seductive power of isolating and abstracting. But perhaps an empathic mathematician would be a *proportionist*, one who re-contextualises and evaluates appropriateness. [19] It is interesting to speculate whether mathematics could be influential in initiating another evolution in human thought, and whether that evolution has a chance of reclaiming Illich's lost sense of proportion. It would seem, perhaps quite ironically given the foregoing discussion, that mathematics, influential as it is in our current world and sensitive as it can be to many different forms of proportion, could be uniquely situated to lead the way.

Melancholy and the 'Mathematical Mind'

> Mathematics is melancholy's mirror-writing.
> (Friedrich Dürrenmatt, 1989, p. 82)

This section takes a relatively and necessarily brief look at a topic, the 'mathematical mind', that has attracted interest and comment down the ages. As the next section indicates (concerning the 'mathematical brain'), this focus may be taking a specifically contemporary turn. At different places in this book, occasional reference has been made to aspects of outlook, temperament, disposition or personality in relation to mathematicians, most particularly in Chapters 4 and 6. A number of other authors have remarked indirectly or somewhat in passing upon such psychological elements too, usually in connection with interviews with mathematicians. [20]

The psycho-historical strand we explore here involves connections between melancholy and the mathematical mind. But why melancholy? If we had to create a comparable set of three 'de-'s to Balacheff's triple, discussed earlier in this chapter, say in an attempt to characterise a modern sense for this somewhat archaic word, it might well be despondency, dejection and delusion or despair. What possible *pleasure* could there be in that? And what particular connection might it have to mathematics? [21]

The melancholy disposition of the mathematical mind

One of the challenges facing writing about this area is the diverse set of ascriptions (despite its central place) that the notion of melancholy has had in more than two millennia of what we would term 'psychological' writing and of literature in general. (For a very helpful and thorough commentary on a historical selection of such 'psychological' writing involving melancholy, see Radden, 2000.) In particular, there has been a remarkable back and forth over the centuries as to whether melancholy is a particular trait or mood, one present in everyone to greater or lesser extent, or whether it refers to a specific mental malady or disorder in need of treatment.

As far back as Aristotle, we find speculations concerning the nature of the creative mind – as well as the important instance of mathematics within such accounts – and the state in which it prepares itself prior to and during creative work: melancholy contemplation. In the pseudo-Aristotelian work *Problems* [22], the question is posed, "Why is it that all men who have become outstanding in [natural] philosophy [...] are melancholic [...]?" (Hett, 1957, p. 155). It is apparent even in this very early account that melancholy is not being offered as a positive trait, but at least it seems one with compensations on occasions, related to exceptional 'creative energy' or 'genius'.

There are two historical periods where this latter connection is particularly explicit: during the early Renaissance and in nineteenth-century Romanticism [23]:

> The glorification of melancholy and the birth of the modern notion
> of genius can be traced to Florentine Neoplatonism, and particularly
> to the work of Marcelo Ficino [...] (Radden, 2000, p. 13)

Ficino, the most significant philosopher-translator in the Renaissance, also claimed melancholy to be endemic among scholars. He dedicated the first part – entitled 'On caring for the health of men of letters' of his *Book of Life* to an account of melancholy, its causes and putative cures. In his extremely lively introduction to Ficino's *Book of Life,* Charles Boer (1980, p. xiii) observes:

> Melancholy, he [Ficino] thought, was a natural condition of the
> soul in the body, and the scholar-philosopher was particularly
> prone to it. [...] For the neo-Platonist, the soul does not want to be
> in the body, and melancholy is its cry for escape.

Part of the tacit suggestion of the previous sub-sections in this chapter is that mathematics in particular invites (as well as provides a means for) the soul to leave the body.

One image that captures some of these elements [24] is Dürer's famed engraving *Melencolia I* (see Figure 1 opposite). For reasons that will become clearer shortly, we have chosen to spend time with interpretation of her facial expression, focusing especially on accounts of her eyes and gaze.

Dürer himself was both melancholic and mathematically inclined:

> In mathematics, above all, to which he had devoted half a lifetime
> of work, Dürer had to learn that it would never give men the sat-
> isfaction they could find in metaphysical and religious revelation,
> and that not even mathematics—or rather mathematics least of
> all—could lead men to the discovery of the absolute, that absolute
> by which, of course, he meant in the first place absolute beauty.
> (Klibansky *et al.*, 1964, p. 364)

Fifteen years after Klibanksky *et al.*'s extensive account of melancholy appeared in English, and while paying tribute to the enormous labour represented in their book, key elements of their interpretation of this engraving were challenged by Frances Yates (1979). In her book, Yates argued for an understanding based more on medical magic and also on Cabbala, what Klibansky *et al.* term *iatromathematics*. Of interest to our very brief account is Yates's interpretation of the sleeping hound as:

> the bodily senses, starved and under severe control [...] the inactiv-
> ity is not representative of failure but of an intense inner vision. The
> Saturnian melancholic has 'taken leave of the senses' [...] (p. 56)

But, for us, what is most striking is the link between a personification of melancholy and mathematics, reflecting the middle term of Martin Luther's ironic triad, "Medicine makes men ill, mathematics sad, theology wicked". But curiously, the parallels here are not exact: the first and third (medicine and theology) embody contrary claims. But we know of nowhere where it is claimed mathematics is *intended* to make one happy.

Earlier authors who had written about this topic include the scholar Henry of Ghent in thirteenth-century Paris (the potent phrase 'the melancholy disposition of the mathematical mind' which titles this sub-section is attributed to him). Henry divided scholars into two differing sorts: those who have direct access to the essence or spirit of things (in fact to the deity) and those, whom he called *mathematicians,* who are obsessed by structure and form which, interestingly in this context, he conceived as embedded in actual matter in physical space. [25]

Of mathematicians, Henry wrote:

> Their intellect cannot free itself from the dictates of their imagination [...] whatever they think of must have extension or, as the geometrical point, occupy a position in space. For this reason such people are melancholy, and are the best mathematicians, but the worst metaphysicians; for they cannot raise their minds above the spatial notions on which mathematics is based. (in Klibansky *et al.*, 1964, p. 338)

Figure 1: Albrecht Dürer, Melencolia I, *1514,* © *The Trustees of The British Museum*

Blindness, solipsism and the 'mathematical mind'

We suggest one aspect of twenty-first-century work on the mathematical aesthetic will involve an ever-deeper analysis of the mathematical psyche, or what for many hundreds of years previously had been referred to and reified as the 'mathematical mind'. Historically, there have been attempts to characterise the 'mathematical mind', usually by comparison or contrast with some other specific kind, e.g. as with Henry of Ghent, the 'metaphysical' mind, rather than simply the 'non-mathematical' mind. Pascal (1659/2000), for instance, in his *Pensées*, contrasts *l'esprit de géométrie* (the mathematical mind) with *l'esprit de finesse* (sometimes rendered into English as the subtle or intuitive mind) in terms of approaching and understanding the world. [26]

The distinction is not that clear, but rests on whether there is a paucity or not of explanatory principles to be drawn on, whether the relevant principles are close or removed from common usage, and whether a judging wisdom and evaluation (and hence the aesthetic) is to be drawn upon. Throughout, his writing is infused with the metaphor of sight and its link to 'sense' in both senses. But it is also clear that, for Pascal at least, mathematicians lack a certain sensibility.

Throughout the subsequent century, particularly but far from exclusively in England and in France, mathematics repeatedly found itself employed in discussions of mental life, including the influence of the five human senses upon it, especially in essentialist arguments about conceptions in and of the mind. One specific context for such exploration was in work with the blind, especially in relation to what became known as Molyneux's Problem (see Degenaar, 1996).

This 'problem' was basically a philosophical thought experiment proposed to John Locke in 1688 (and published in his *Essay Concerning Human Understanding* in 1690) involving the sensory roots of ideas versus the possibility of innate ideas. The problem queried whether persons born blind would, if they were to regain their sight, be able to recognise objects visually. Specifically, would he or she be able to distinguish a cube from a sphere by sight alone when sight was restored. (Our brief account here is directly based upon Jessica Riskin's (2002) insightful chapter, 'The blind and the mathematically inclined', which she opens with this 1751 remark of Voltaire's, "Mathematics [… is] the staff of the blind", p. 19.)

Leibniz argued in 1705 against Locke:

> Leibniz distinguished "images," specific to the senses, from "exact ideas, which consist of definitions." [...] Though each image belonged to an individual sense, the composite definition-idea belonged to "the common sense, that is to say, the mind itself." Geometry dealt in ideas not images. So a blind man using tactile images and a paralytic using only visual images would arrive at the same geometry, consisting of the same ideas. (Riskin, 2002, p. 24)

The key progress made quite early in the eighteenth century in accounts of Molyneux's problem was due to the development of cataract surgery which resulted in the restoration of sight to some individuals, thereby allowing the prospect of an empirical rather than an epistemological resolution. Riskin observes of Julien de La Mettrie's later view, "Geometry might enter the soul equally through the eyes or the tips of the fingers" (p. 43).

The philosophical context of this work involved a polarisation around 'sensibility', a key term in Riskin's significant eighteenth-century study of Western European science and culture [27]. Sensibility meant:

> a physical sensory receptiveness to the world outside oneself, whose consequence was emotional and moral openness. Its opposite was a physical insensitivity that brought solipsism. (p. 21)

Mid century, Etienne de Condillac argued forcefully that one could have sensation without sensibility and that the source of sensibility was to be found in touch and only in touch, in finding something outside of oneself that did not return an inner sensation of touch when touched. (See footnote [10], in respect to which Riskin's chapter has a great deal to add, not least the confusion during the eighteenth century as to whether sight or touch were the primary sense of the intellect, of the experimental, of the rational, of the true.) Riskin summarises:

> one could perceive the external world but fail to perceive it *as* external. In that case, one would be a being for whom nothing but oneself existed. Insensibility meant isolation from "the commerce of others"; it meant solipsism. (p. 52; *italics in original*)

A key figure in such discussions was that of Denis Diderot (also discussed in Chapter 9) and, in particular, his 1749 *Lettre sur les Aveugles*:

> Drawing upon [... among other sources] a memoir on the life of the blind Nicolas Saunderson, late Lucasian Professor of Mathematics at Cambridge, Diderot argued that the blind, because of their impoverished sensibilities, turned their minds inward and tended to think in abstractions. This made them natural mathematicians [...] a blind man's view of the world was made of geometrical abstractions. [...] Diderot argued first that the blind because of their sensory deprivation, were necessarily abstract mathematical thinkers; and second that abstract mathematical thinking amounted to emotional and moral solipsism. (pp. 21, 22, 53)

Riskin goes on to document how the perception of an overlap between the blind and the mathematically minded (again, arguably a compensation or due to a heightened inner awareness) was a virtual commonplace during the eighteenth century. (In passing, she also reports how, in an 1801 text, Pierre Simon Ballanche identifies the central emotional quality of blind poets (especially Homer and Milton) as being that of melancholy.) In his *Letter about the Blind*, Diderot cited mathematician Mélanie de Salignac's declaration, in terms that closely echo our earlier discussion of melancholy, that the mathematician "spends almost all his life with his eyes closed" and that mathematics "was the true science of the blind".

Autism and the 'Mathematical Brain'

In her attempt to bring up to date her historical account of the fluid meanings for and associations with the notion of melancholy, Jennifer Radden (2000) sensitively raises the question of whether the contemporary category of 'clinical depression' can be seamlessly grafted onto 'melancholy'. Some medical authors have tried to do this, as if the two were always and everywhere the same thing. Here, we wish not so much to pursue this specific line as indicate a different disorder, Asperger syndrome, as being of possible interest and relevance to the 'creative genius in mathematics' strand of the foregoing discussion. In addition, it offers a more contemporary version of the 'mathematical mind' discussions (albeit one which resolutely returns to the cognitive and even to the architecture of the brain), with specific links both to melancholy and the blind.

Sixty years ago, paediatrician Hans Asperger published a seminal paper on 'autistic' children (from the Greek *autos*, meaning "self", individuals for whom in some important sense there is only themselves or perhaps no one). In it, he characterised the essential disturbance of autism as one of contact, "of the lively relationship with the whole environment" (1944/1991, p. 74). In relation to abstraction, he claimed, "In the autistic person, abstraction is so highly developed that the relationship to the concrete, to objects and to people has been largely lost [...]" (p. 85).

Asperger himself noted a link between mathematical ability and some of the individuals he was seeing in his practice (who might retrospectively be characterised as mildly autistic but very able in highly specific realms):

> We have seen that autistic individuals [...] can [...] achieve professional success, usually in highly specialised academic professions, often in very high positions, with a preference for abstract content. We found a large number of people whose mathematical ability determines their professions: mathematicians, technologists [...]. (p. 89)

As he noted, one characteristic feature of a person with autism is in the nature of language use:

> Autistic language is not directed to the addressee but is often spoken as if into empty space. This is exactly the same as with autistic eye gaze which, instead of homing in on the gaze of the partner, glides by him. (p. 70)

It will not have escaped the attentive reader that the first sentence provides a pretty fair description of mathematical text: mathematics as autism's mirror-writing perhaps? Mathematician Brian Rotman has focused specifically the non-subject nature of the one addressed by mathematical discourse:

> Mathematical addressees are theoretical and impersonal: mathematicians prohibit their codes from making any sort of reference

> to the individual characteristic of the reader; or to his subjectivity or
> to his physical presence in the world. (in Walkerdine, 1988, p. 186)

Another linguistic feature that is often disturbed in individuals with autism is the use of pronouns, in particular employing 'I' for 'you' and 'you' for 'I' (see, for example, Frith, 2003, pp. 124-125). This is also the case with other pairs of terms that tie language to context, such as 'this' and 'that' and 'here' and 'there'. Speculation about possible causes for and consequences of such deictic confusion in autistic individuals regarding the positions and points of view of speaker and listener has been widespread. Once again, turning to mathematical text and in particular the means by which 'universality' is inscribed within it, we find that such features are actually absent:

> everything else is excluded, including the addresser and addressee – there is no 'I' and 'you' in a mathematical string. There is no grammatical subject or object. For Lacan it is the pronouns I, you, he/she/it which position the speaking subject 'in language' and thereby fix an identity. It is in all these senses that I have wanted to suggest that much is suppressed in order to reach the mathematical string itself. (Walkerdine, 1988, p. 199)

And we wish to suggest, in light of the above, that individuals with Asperger syndrome perhaps have less to suppress or possibly that the necessary suppressions to gain mathematical fluency may cost them less in psychic terms.

As indicated above, one of the characteristic disturbances of autistic spectrum disorders has to do with fleeting gaze (whether at people or things) and indifferent contact (especially social) with the outside world, the world of context:

> One can never be sure whether the glance goes into the far distance or is turned inwards [...] (Asperger, 1944/1991, p. 69)

There is a striking similarity between this and Poincaré's observation about the analyst Charles Hermite, he "whose eyes 'seem to shun contact with the world' and who seeks 'within, not without, the vision of truth'" (in Hadamard, 1945, p. 109), as well as with the above discussion of psychological projections onto physical blindness. And perhaps unawarely recasting Aristotle's question XXX from the *Problems*, Asperger suggested:

> It seems that for success in science or art a dash of autism is essential. For success the necessary ingredient may be an ability to turn away from the everyday world [...] (in James, 2003, p. 63)

The children and adults with whom Asperger, Kanner and others since have worked exhibit what has come to be termed 'autistic aloneness', a notion which has links back to above eighteenth-century concerns with solipsism. And significant depression is a common concomitant of autism. We are certainly not suggesting, even among a selected 'genius' category of 'historical great minds', that those who were seen as melancholics in earlier times would all today have been diagnosed with autism or specifically Asperger syndrome.

Nonetheless, it is possible to see at least an overlap among the various topics of the last section in terms of speculation about the 'mathematical mind'.

The connection between eminent mathematicians and autism has been alluded to, both from within the community of autism specialists and by professional mathematicians. Specifically, within the past five years psychologist Michael Fitzgerald and mathematician Ioan James have separately written challengingly on this theme. Fitzgerald (2000, 2002), in two letters to the editor of the *Journal of Autism and Developmental Disorders* which draw on biographical material in relation to contemporary diagnostic criteria, raises the question whether Asperger syndrome individuals' cognitive style is also a "mathematical style"?

While the evidence offered in these short communications is slight, it does give the feel for how a far more extensive psycho-biographic argument might go with respect to such retrospective diagnosis, as well as providing a list of potentially relevant individuals (including Cauchy, Erdös, Galois, Gauss, Hardy, Lagrange, Lobachevsky and Riemann). More recently, Fitzgerald (2004) has published a book, *Autism and Creativity: Is There a Link between Autism in Men and Exceptional Ability?*, in which he provides a more detailed (though still to our minds speculative) examination of half a dozen individuals, including Wittgenstein and Ramanujan. [28]

James (2002, 2004) has written two volumes of biographical profiles of mathematicians and physicists, though no mention within them is made of Asperger syndrome. However, in November 2003, he published an article in *The Mathematical Intelligencer* entitled 'Autism in mathematicians'. In it, he discusses a range of individuals as well as characteristic traits of the disorder. He also speculates on the effects of myopia, exemplified by Lie, Poincaré, Levi-Cività and Noether, mentioning its possible genetic link to autism.

Lastly, James asks:

> Why are mathematicians, along with computer scientists, commonly regarded as loners and placed in a group with geeks and nerds? Could it be that the type of personality which inclines people towards mathematics has something to do with this? And could it also be that here is part of the explanation for the difference in the relative number of men and women to be found in mathematics? (p. 64)

In *The Essential Difference: Men, Women and the Extreme Male Brain,* Simon Baron-Cohen (2003), head of the autism research centre at Cambridge University, argues that what he terms *systemizing* ("the drive to analyse, explore and construct a system", p. 3) and *empathizing* ("the drive to identify another person's emotions and thoughts, and to respond to them with an appropriate emotion", p. 2) comprise two core human mental processing traits linked to attention. [29] And these traits seem relatively independent of one another, with mathematical ability identified as "one of the clearest examples of systemizing" (p. 116).

However, according to Baron-Cohen, their distribution in the population is gender linked: statistically more males are higher systemizing than females and more females are higher empathising than males (see also Baron-Cohen and Wheelwright, 2004). What he unhelpfully terms the 'male' brain is the combination of higher systemizing than empathising (measured in units of standard deviations of the two measures) and the 'female' brain the reverse.

Baron-Cohen's most striking proposal relates to "extreme" examples of the 'male' brain (very high system, very low empathy). He links this directly to various forms of autism (especially Asperger syndrome) and exemplifies such a brain by means of the Cambridge mathematician Richard Borcherds – a Fields medal winner. Baron-Cohen writes of him, "His talents in mathematics have resulted in his finding a niche where he can excel (to put it mildly), and where his social oddness is tolerated" (p. 163).

Baron-Cohen has proposed a diagnostic empathy measure (the 'reading the mind in the eyes' test), reflecting the diminished capacity of individuals with autism to 'read' emotion from photographs of human pairs of eyes. From at least the time of Albrecht Dürer onwards, there has been a significant link made between human gaze and mathematical minds. One of the small ironies around discussions of the significance of Melancholy's in-turned gaze, in particular whether it be sad or pleasurably entranced from within, is that many Asperger individuals would be unlikely to be able to discriminate among suggested possible states or even attribute a specific mental state to her.

In the last two sections, we have seen mathematics characterised as the melancholy science, the blind science and now, perhaps, the autistic science. While not doing so explicitly, in the back of our minds we have been holding up Balacheff's triple-detachment list to the discussions and theorising of different writers through the ages: decontextualisation, depersonalisation, detemporalisation. The first two are without question, features of autistic discourse; as for the last, we are simply unaware of as yet, whether there is a 'preference' for the timeless present.

Both Fitzgerald and James write of the strong connection with intense originality and creativity in some very successful individuals with Asperger syndrome, and both offer suggestions as to why such individuals might be found at the very pinnacle of their disciplines. We, here, are caught by some challenging questions, questions we simply pose to move towards a conclusion of this section.

What draws people into mathematics? Does mathematics necessarily demand extreme abstraction and, if so, what are the costs and consequences for empathic human beings? Is this an instance where overly-ordered, overly-detached individuals nevertheless find in mathematics sufficient characteristics in common to support their own psychology? Is mathematics, in fact, an important component of some individuals remaining sane? Lastly, is there also a viable question about a possible reversible influence, asking how mathematics is shaped by the characteristics and qualities of the individuals who do it, as well as asking how mathematics shapes those who undertake it.

Mind, brain and mathematics

Other contemporary talk of the mathematical 'brain' (used in preference to the mathematical 'mind' [30]) can be found in the extensive exchange between Jean-Pierre Changeux and Alain Connes (1995), *Conversations on Mind, Matter, and Mathematics,* mentioned at the outset of this chapter. While often finding themselves disagreeing, whether about the nature of mathematical objects or the possibility of machine replication of mathematical activity, it is engaging to watch each trying to undergird their own and undermine the other's account, rather in the manner of wrestlers.

Changeux puts atoms and neurons at the explanatory heart of constitutive generative enquiry and creativity in mathematics ("Mathematical objects exist solely in the mind of the mathematician, not in some platonic world independent of matter. They exist in the neurons and synapses of the mathematicians who produces them", pp. 11-12), while Connes resolutely counters with his purely descriptive, discovery role for the brain (p. 22). What the latter terms 'archaic mathematical reality' for him pre-exists any human mental activity in a completely independent manner. We are back very firmly in the arena we described in Chapter α. But there is also a stimulating discussion of illumination.

In discussing the possibility of Darwinian natural selection operating on (potential) mathematical objects, the conversation runs as follows:

> *Connes:* Illumination is not only marked by the pleasure—the exhiliration!—one inevitably experiences at the moment it strikes, but also by the relief one suddenly feels at seeing a fog abruptly lift, and disappear. [...]

> *Changeux:* You make me think of the mystical ecstasy of Saint Teresa of Avila.

> *Connes:* Mystical ecstasy must certainly excite the same regions of the brain—aesthetic harmony as well—but for other reasons I should think. (pp. 147-148)

Tahta (1996), in his essay review of this book, sees the still centre of this genuine exchange of views being the possibility of mathematics serving as mediating third between mind and matter, in the same way that 'transitional objects' mediate between fantasy and reality for the very young child. In so doing, he was echoing John Dee's comment over four hundred years earlier, one that we quoted in Chapter α:

> For, [*Things Mathematicall*], being (in a manner) middle, between things supernaturall and naturall: are not so absolute and excellent as things supernaturall; Nor yet so base and grosse, as things naturall: But are things immateriall, and nevertthelesse, by material things able somewhat to be signified.

Psyche in the Realm of Essence

> He said to me: "It is done. I am the Alpha and the Omega,
> the Beginning and the End". (Revelation, 21:6)

Below is the most infamous extract from Gottlob Frege's (1884/1978) book *The Foundations of Arithmetic*, one that could actually serve as a trenchant summary of his anti-psychological views in relation to mathematics and its origins (and the apparent fears they evoked in him, "But this account makes everything subjective, and [...] does away with truth").

> What is known as the history of concepts is really a history either of our knowledge of concepts or the meaning of words. Often it is only after immense intellectual effort, which may have continued over centuries, that humanity at last succeeds in achieving knowledge of a concept in its pure form, in stripping off the irrelevant accretions which veil it from the eyes of the mind. What, then, are we to say of those who, instead of advancing this work where it is not yet completed, despise it, and betake themselves to the nursery, or bury themselves in the remotest conceivable periods of human evolution, there to discover, like JOHN STUART MILL, some gingerbread or pebble arithmetic! [...] As far as mathematicians are concerned, an attack on such views would indeed scarcely have been necessary; but [...] for the philosophers [...] I found myself forced to enter a little into psychology, if only to repel the invasion of mathematics. (pp. VIIᵉ-VIIIᵉ)

As will be readily apparent by now, we hold no such reservations about psychology suitably conceived. Should we need support, we simply recall the quotation from Henri Lebesgue, cited at the outset of Chapter 8, which contradicts Frege by necessarily linking psychology (and 'even the aesthetic') to the foundations of mathematics. [31]

Given this chapter's location within our book, as well as our occasional glancing back to the initial chapter, we find it interesting in closing to try to narrow down somewhat what might be meant by one of the central terms of the book, namely 'mathematics' or 'the mathematical'. (We do this bearing in mind that, at least as much as 'melancholy' if not more so, neither of the terms 'mathematics' nor 'mathematician' have had stable or firm meanings – nor even constant resonance – across the ages.) To this end, we simply propose to point to a series of boundaries, which we shall locate by means of intentionally naïve questions.

(a) Our first question concerns the human/machine interface. are we prepared to call something that inanimate machines do 'mathematics'? René Thom (1973, p. 205) has characterised mathematics as "the science of the simulation of automatisms", though it is clear from his discussion of the role of the human unconscious in this piece, that the automatisms he had in mind were human ones. Nevertheless, as we saw in Chapter 7 and elsewhere, machines are increasingly being used in the service of such simulation.

We have heard colleagues denigrating the mathematical results of computer theorem-provers (for instance, Jürgen Richter-Gebert's comment 'stupid proofs of simple theorems'), though there are now instances of computers generating not so-simple proofs. What these colleagues may be pointing to is the fact that the mathematician must interpret the results in the end, no matter how many computations are made, irrespective of how sophisticated they may be. As David Henderson and Daina Taimina observed in the third chapter in this book: "The goal is to understand meanings" (p. 59).

That said, a quick glance at Jonathan Borwein's chapter in this book suggests that neat distinctions between human thinking and human tools are misleading. We can no longer do without those inanimate machines, which not only provide results, insight, visualisations, counter-examples and verifications for many pieces of mathematics, but which markedly facilitate the whole enterprise of communication and publication within the mathematical community. If, in 2005, we still feel the need to separate the work of the machine from that of the mathematician (as suggested by the upcoming split publication of a purported proof of Kepler's conjecture), then the question may be, in the end, how prepared we are to dissolve the human/machine interface.

(b) Our second question is broadly historical: what are the earliest instances of human activity we are willing to accept as mathematics (i.e. where do we wish to locate the beginnings of mathematics and why). [32] In light of this, what could be considered the earliest mathematical artifact (and how would we know)? What do different answers reveal about the investments of different proposers? What has mathematics been?

For example, the interesting controversies around notched wolf bones some ten thousand years old found in what is now Zaire (Fauvel and Gray, 1987, p. 5) or neolithic stone balls (Tahta, 1980, as well as in his Chapter 9 of this book) have to do with their contested status of whether or not they are to be taken as inherently mathematical objects. Such discussions involve the usual problem of inferring intention and design from physical attributes alone. But the clock for mathematics keeps being restarted, sometimes in unexpected directions. For some, the real numbers were only brought into existence during the second half of the nineteenth century with the publication of Dedekind's work: he himself tells us so in his *Was sind und was sollen die Zahlen?* One of the more recent Year Zero attempts locates origins in France during the 1930s and 1940s with the Bourbaki group's encyclopaedic and abstractionist goals.

We have been particularly interested in the recent scholarship around cave art in Europe and North America. For instance, David Lewis-Williams (2002) argues that this Upper Palaeolithic art marks not only the origins of art itself, but also of religion and consciousness. That stone-age *trivium* is tantalising, particularly given the mathematics–art relationships discussed by Martin Schiralli in Chapter 5 and the mathematics–religion connections that, somewhat to our surprise, became a central theme in several of the chapters, particularly in Chapter 8 and 9.

Finally, we are also reminded of the mathematician André Weil's (1992) description of the mathematical experience, which hints at its other-worldliness, as "the state of lucid exaltation in which one thought succeeds another as if miraculously, and in which the unconscious seems to play a role" (p. 27). In the end, this question may harken back to William James (1902): why do mathematics, religion and art all seem to give rise to extreme, integrative experiences?

(c) The third question refers to the chronological present and relates to the distribution of mathematics, mathematical awareness and sensibility within cultures, as well as globally across cultures. In other words, this relates to what we could call the *ethnomathematical* question: in what sense does everyone mathematise or 'do' mathematics to some extent? [33]

Is it true that everybody counts (as the eponymous 1989 US National Academy of Sciences' report claims)? Is it right, as Alan Bishop (1991) has argued, that there are at least six (his total) mathematico-cultural activities that all human groups engage in (including counting, locating and playing) that lead to mathematics? Or shall we use Saunders Mac Lane's (1986) longer list, as was briefly discussed at the end of Chapter 6? Perhaps they are all to be seen simply as varied instances of *proto*mathematical activity (as Yves Chevallard, 1990, and others have used this term), namely something that is not yet mathematics, but that can, albeit under certain stringent circumstances, lead to it.

While accepting the polygenesis of protomathematical experiences, Chevallard wryly warns against too facile a presumption of continuity between everyday and mathematical perceptions, awarenesess and experiences (which, seen in a more generous light, might have been part of Frege's intent). Writing of a mathematics education turn toward the cultural more than a decade ago, Chevallard notes:

> It is, I observe, a desperate attempt to prove that mathematics is *not foreign* to the child's everyday experience. (1990, p. 6; *italics in original*)

This moves us back to *mathematisation* as a human activity, "where something not obviously mathematical is being converted into something which obviously is" (Wheeler, 1982, p. 47). Finally, what are we to make of author Russell Hoban (1982) who, in his post-apocalyptic novel *Riddley Walker*, had a character claim: "Them as counts counts moren them as dont count" (p. 18)?

(d) Our fourth and final question concerns the genesis of mathematics, not historically within cultures, but psychologically within the individual. Are all mentally functioning humans born with the possibility for mathematics? Does the potential for mathematisation become actualised, say, in a manner akin to language acquisition: that is, only with adequate external input? What necessary adult awareness needs to be brought to bear on a young child in order to draw attention to the very possibility of mathematics? The work of Caleb Gattegno (e.g. 1970, 1998) is a touchstone here, not least with his concentrated and fertile attention to the mathematical powers of infants made evident by and inherent in their learning of their mother tongue.

> To stress and ignore *is* the power of abstraction that we as children use all the time, spontaneously and not on demand, though in its future use we may learn to call it forth by demand. And teachers insist that we teach abstraction to children through mathematics at the age of twelve! (1970, p. 28; *italics in original*)

In Chapter 6, William Higginson presented an anecdote about an individual being sensitive to mathematical–aesthetic issues at a very young age, while the following quotation from mathematician Michael Sipser (1986) provides another:

> I was very young […] and I remember my father folded down the flaps of a cardboard box so that each was holding down the next. And I remember [...] being amazed, it was so perfect [...] my first experience of joy in abstract thought. (p. 80)

In the end, a core pair of questions to take into an emerging twenty-first century might well be these. What are the psychic gains and losses in doing mathematics? Why are some so willing to engage and persist with it, while others are equally resolute in their abiding refusal?

Notes

[1] In *Symposium,* in response to the question "Why does the person who loves love beautiful things?", Plato had Socrates reply, "To posses them for himself" (1998, p. 85). Posession is both the rationale for and the reward of love (*erôs*).

[2] This apparently simple title is not grammatically straightforward. Due to the word 'inconscient' ("unconscious") being both an adjective and a noun in French (as in English), 'l'inconscient mathématicien' could be variously rendered as "the unaware/ unconscious mathematician" and "the mathematician unconscious". This latter phrase comprises both the unconscious of the mathematician and the mathematical part of the unconscious. However, the first reading requires going against a con- ventional order of noun followed by adjective in French, while the latter requires the noun 'mathématicien' being treated as an adjective. As a Lacanian, Bouleau likely enjoys playing with all possible significations.

Right at the end of *Psychoanalytic Politics,* Sherry Turkle's (1978) partial history of Lacanian psychoanalysis in France, Lacan's deep involvement with mathematics, especially knot theory, is described. Indeed, what turned out to be his final project before he died was to explore the Paris asylum records of Georg Cantor (see also Charraud, 1994, 1997). In terms that prefigure one central focus of this chapter, Turkle concluded:

> For Lacan, mathematics is not disembodied knowledge. It is con- stantly in touch with its roots in the unconscious. This contact has two consequences: first, that mathematical creativity draws on the unconscious, and second, that mathematics repays its debt by giving us a window back to the unconscious. (p. 247)

[3] Mention of the theme of creativity takes us back to Chapter α, where we identified it as one of three primary foci of attention explored by various writers on the mathematical aesthetic over the previous hundred years. The first specific focus, on the nature of mathematical enquiry, touched on the general millennia-old debate fuelled by Plato concerning creation versus discovery. And *mathematical* creation has always played a central role in such discussions. Dick Tahta argued in Chapter 9 that one central feature is that creation *and* discovery are both strongly linked to the early psychic development of the individual, namely the coming to trust and believe in things (including human beings as things). And the contribution of the aesthetic in relation to mathematical creativity, not least in regard to giving rise to what Poincaré (1908, p. 20) termed 'the selected fact', is perhaps its most significant role.

[4] Instances concerned with aesthetics and intuition, as well as a wider range of feelings associated with mathematics, can be found in Chapter 5 of Burton (2004), as well as in Albers *et al.* (1985, 1990), Bouleau (1997) and Sfard (1994). Benjamin Bloom's extensive study during the 1930s of the personal and family history of North American prodigies in a variety of fields (including mathematics) identified a number of characteristic personality traits. Some that were identified in the mathematical subjects included a "penchant for solitude" and a "desire for precision" (see Gustin, 1985), as well as being independent-minded. The mathematicians made frequent references to enjoying being able to "derive from scratch", as well as what could be called a fundationalist tendency, a desire to get to the 'bottom' of things (evoking for us Charles Murray's seeker of "the inner truth of things" – see Chapter 6). But, as Roland Barthes (1975) reminds us in *The Pleasure of the Text,* such pleasure is Oedipal: the desire "to know, to learn the origin and the end" (p. 10).

[5] Nevertheless, Ken Ribet (in Singh, 1998, p. 288), when faced with being wrong about his claimed proof of the link between the Taniyama–Shimura conjecture and Fermat's last Theorem, observed 'I had an immediate terror'. Philosopher of logic Ross Skelton (1993), in an article on mathematical problem solving and Wilfred Bion (a follower of psychoanalyst Melanie Klein), has suggested that the creative process would involve a return to the Kleinian 'paranoid–schizoid' position. This would be more chaotic and potentially destructive than what we see Poincaré as trying to describe. And Nicolas Bouleau (2002), writing in the context of Galois's personality and drawing on Lacan's view of creativity as closely linked to paranoia and his seeing of 'science as a successful paranoia', remarks:

> I want to acknowledge here a personal experience corroborated by other mathematicians: certain phases of mathematical research have plunged me into a state of dread and emotional fragility akin to accounts of pathological cases of paranoia or paraphrenia. (p. 178)

[6] Nicholas Jackiw, in his discussion in Chapter 7 of the aesthetic motivations for dynamic geometry activity, identifies the dual aesthetic intrinsic to dynamic geometry, and perhaps to mathematics. On the one hand, there is the detached purity and apparent other-worldliness of mathematical objects, while, on the other lies the felt experience that comes from tangibly manipulating mathematics towards one's own passionate and personal need for understanding and connection.

[7] Disinterested pleasure is possible because the subject forms a representation of the object. Therefore, argued Kant, if contemplation of a mental representation causes pleasure due solely to the way it stimulates a subject's cognitive faculties

(rather than due to stimulation of the subject's desires or interest), every other subject will be able to take the same pleasure in this contemplation. Kant thus held that to judge something to be beautiful is to make a judgement that applies universally; one is 'judging' that everyone ought to find the judged object beautiful as well.

[8] For example, Beardsley's notion of "detached affect" certainly implies some emotional separation: "a sense that the objects on which interest is concentrated are set a little at a distance emotionally" (p. 288).

[9] Art critic Meyer Schapiro (1946/1978), writing of Vincent van Gogh in contrast to an Impressionist painter's lesser concern with the object, commented powerfully on his *attachment* to the object, what Schapiro termed his "personal realism":

> I do not mean realism in the repugnant narrow sense that it has acquired today, [...] but rather the sentiment that external reality is an object of strong desire or need [...] (p. 93)

[10] Footnote 19 of Chapter 8 made mention of a mathematical contrast between the senses of sight and sound. The frontispiece to Gabriel Josipovici's (1996) striking book *Touch* (a book which goes further, contrasting the senses of sight and touch) asserts that:

> it is possible to feel comfortable in the world and in our relationships with others only if we value touch over sight, if we respect distance but also work to overcome it [...] although sight seems to give us the totality of what we behold, it is only when we walk or feel our way across the distances that things become more than images and begin to constitute the world in which we, as touchers and not mere observers, are included.

Nicholas Jackiw's observations in Chapter 7 about the haptic as well as visual stimuli evoked in working with *The Geometer's Sketchpad* are particularly pertinent here.

[11] Yet, as we have noted earlier in this chapter, in mathematics there is also an extremely strong pressure towards 'perfect detachment', namely a complete ego-suppression – see also Rotman (1988). However, Hermann Weyl (1949, p. 75) once noted that a trace of the subject position of the mathematician can still be located in space by means of the origin: "The objectification, by elimination of the ego and its immediate life of intuition, does not fully succeed, and the coordinate system remains as the necessary residue of the ego-extinction."

[12] There are fascinating links here to jurisprudence and to Oliver Wendell Holmes's challenging declaration in *Lochner v. New York* that, "General principles do not decide concrete cases" (in Menand, 2002, p. 34). "It was Holmes's genius as a philosopher to see that the law has no essential aspect" (p. 35). Santayana (who was quoted in Chapter 1) referred to mathematics as "the realm of essence".

[13]   Perfection is terrible, it cannot have children.
        Cold as snow breath, it tamps the womb
        (Sylvia Plath, 1965, p. 74)

Catherine Chevalley echoes Plath's observation when remembering the intrusion Bourbaki as a group made into her life as a girl, claiming what made it worse for her was that this was a group who "reproduced without a single woman" (in Chouchan, 1995, p. 38).

[14] Although the quest for certainty seems a mainstay of mathematics, in Chapter 4 Nathalie Sinclair noted Wolfgang Krull's (1930/1987) insinuation that some mathematical pursuits exaggerate the quest, at the expense, perhaps, of aesthetics. Krull expressed the view that those attracted to the study of foundations are the least aesthetically oriented mathematicians, since they are "concerned above all with the irrefutable certainty" (p. 50) of their results. Yet links between beauty, pleasure and certainty are still commonly remarked upon. As just one example, Joseph Liouville, talking of his work to ready some of Galois's final papers for publication, noted, "I experienced an intense pleasure at the moment when, having filled in some slight gaps, I saw the complete correctness of the method by which Galois proves, in particular, this beautiful theorem" (in Singh, 1998, p. 249).

[15] In a section entitled 'The psychoanalysis of mathematics', Connes remarks how when investigating simple but generative mathematical concepts, "one truly has the impression of exploring a world step by step—and of connecting up the steps so well, so coherently, that one knows it has been entirely explored. How could one not feel that such a world has an independent existence?". To this, Changeux interjects "'Feel', you say? [...] I fear the 'feeling' you have of 'discovering' this wholly platonic 'reality' amounts to nothing more than a purely introspective—and therefore subjective—analysis of the problem" (Changeux and Connes, 1995, pp. 30-32).

[16] Rotman (1993, 2000) has produced two books which indirectly and directly explore this theme further, in his pursuit of a successful *semiotic* account of mathematics. His account includes three related 'figures' at work – Person, Subject and Agent – and he refers to the passage between them as "abstractions" (and which we might see as detachments). Rotman comments:

> It would have been more precise to have spoken of forms of principled "forgetting" to describe the processes of reduction and truncation whereby the Person (metalingual, indexical/reflexive, involved in arguments via metasigns) gives rise to the Subject (lingual, collaborative, involved in deductions with signs), who in turn produces the Agent (sublingual, mechanical, involved in actions on signifiers). Baldly, the move from Person to Subject is organized around the forgetting of indexicality, and the move from Subject to Agent around the forgetting of sense and meaning. (1993, p. 91)

When thinking of this sort of "principled amnesia" as a certain form of abstraction, Walkerdine (1988) reminds us that, "For Freud, forgetting is an act of the unconscious" (p. 189).

[17] What possible role can there be for the human in mathematical knowledge, when, as Shapin and Schaffer (1985) simply declare, "To identify the role of human agency in the making of an item of knowledge is to identify the possibility of it being otherwise" (p. 23).

[18] It is striking to juxtapose the foregoing with the following two quotations, from earlier times. Both relate to human knowledge and the possibility of perfect and certain understanding *vis-à-vis* the divine, the customary locus of the perfect and the certain. But the two views expressed below differ in terms of the qualitative similarity or difference of such understanding in relation to the quantity of mathematical truths.

The first comes from Galileo's (1632/1953) *Dialogue Concerning the Two Chief World Systems:*

> But taking man's understanding *intensively,* in so far as this term denotes understanding some proposition perfectly, I say that the human intellect does understand some of them perfectly, and thus in these it has as much absolute certainty as Nature itself has. Of such are the mathematical sciences alone; that is, geometry and arithmetic, in which the Divine intellect indeed knows infinitely more propositions, since it knows all. But with regard to those few which the human intellect does understand, I believe that its knowledge equals the Divine in objective certainty, for here it succeeds in understanding necessity, beyond which there can be no greater sureness. (Salviati speaking, p. 103; *italics in original*)

The second was written by Benjamin Peirce (1850) at the end of a paper he wrote on fractions which occur in phyllotaxis (the formal arrangement of leaves on a stem):

> May I close with the remark, that the object of geometry in all its measuring and computing, is to ascertain with exactness the plan of the great Geometer, to penetrate the veil of material forms, and disclose the thoughts which lie beneath them? When our researches are successful, and when a generous and heaven-eyed inspiration has elevated us above humanity, and raised us triumphantly into the very presence, as it were, of the divine intellect, how instantly and entirely are human pride and vanity repressed, and, by a single glance at the glories of the infinite mind, are we humbled to the dust. (p. 34)

[19] The work of Sherry Turkle and Seymour Papert (1992) has provided some glimpse of how empathic individuals might continue to participate in the culture of mathematics. Influenced primarily by computer programming, rather than mathematics, they have called for an epistemological pluralism that would challenge the "hegemony of the abstract, formal, and logical as the privileged canon in scientific thought" (p. 3). Turkle and Papert have proposed a "revaluation of the concrete" and identify the computer – which stands between the world of formal systems and that of physical things – as a promising and powerful ally. Despite its prevailing image as a logical machine, they see the computer as having the ability to make the abstract concrete, and hence to provide visible, almost tangible, access to mathematical ideas (an observation that Nicholas Jackiw also made in Chapter 7). Their empathic approach (in Worringer's sense) to programming involves a more "flexible and non-hierarchical style, open to the experience of a close connection with the object of study" (p. 9). Using the computer involves insisting on negotiation, relationship and attachment. An empathic approach to mathematics might similarly involve the computer, but in a far deeper way than that ever imagined by educational technologists.

[20] Physicist Helenka Przsiezniak offered a characteristic of physicists: arrogance. "You want to prove that something is right if you believe in it. That's just how it works when you're discussing the 'truth'" (in Baron-Cohen, 2003, p. 65). We find a similar generalisation expressed by Paul André Meyer, in interview with Nicolas Bouleau (1997), when asked whether mathematics perhaps engenders a certain

forme d'esprit ("kind of mind"): "In any case, [it involves] a certain intellectual behaviour, that one could call audacious, if one wanted to be laudatory, presumptuous otherwise" (p. 47).

[21] Here is a short extract from an interview with Andrew Wiles (in Singh, 1998, p. 306):

> People have told me that I've taken away their problem, and asked
> if I could give them something else. There is a sense of melancholy.
> We've lost something that's been with us for so long [...]

Wiles also remarked upon his personal sense of loss (p. 331), as well as his ambivalence: "I got so wrapped up in the problem that I really felt I had it all to myself, but now I was letting go. There was a feeling that I was giving up a part of me" (p. 271).

In 1917, Freud wrote a significant essay entitled 'Mourning and melancholia', where he explored parallels between these two notions. In so doing, he emphasised *loss* in connection with melancholy: not just actual loss of a loved person, but also perhaps "the loss of some abstraction that has taken the place of one" (1917/1955, p. 243). Freud also observed that the melancholic "has a keener eye for the truth than other people who are not melancholic" (p. 246). There is no space here to detail certain links we might under-hear between Wiles's comments about his and others' experience and Freud's account. An open and significant question for us is the nature of the psychic connection between mathematics and loss, as well as the latter notion's link with melancholy.

[22] In the introduction to his translation of the *Problems*, Hett (1957, p. vii) informs us that Aristotle is definitely not the author of this work, at least as it has come down to us, but nevertheless many ancient authors cite problems as being Aristotelian that appear in this work. In particular, Hett argues, "for example, Book XXX. 1, the important problem dealing with the 'melancholic' temperament, is vouched for as Aristotelian both by Plutarch and Cicero" (p. vii).

[23] In literature, there are two periods in English poetry when the 'pleasing melancholy' (a juxtaposition not used ironically) was quite a common allusion: Elizabethan and Romantic verse. John Keats, for instance, begins the final stanza to his *Ode on Melancholy*:

> She [Melancholy] dwells with Beauty – Beauty that must die.

And in the closing lines of *Il Penseroso*, wherein the poet hails 'divinest Melancholy', even John Milton declares himself willing, in older age, to pay her price for granting the intellectual joys to come:

> These pleasures, Melancholy, give,
> And I with thee will choose to live.

Of particular relevance to this chapter, Radden (2000) writes:

> The link with genius was also revived in the literary movement of
> the late eighteenth and early nineteenth centuries. Again the suf-
> fering of melancholy was associated with greatness; again it was
> idealized, as inherently valuable and even pleasurable, although
> dark and painful. The melancholy man was one who felt more
> deeply, saw more clearly, and came closer to the sublime than
> ordinary mortals. (p. 15)

[24] In their extensive historical account of the notion of melancholy, *Saturn and Melancholy,* Klibansky, Panofsky and Saxl (1964) document the ancient tension around melancholy as a 'natural temperament' and a 'pathological disease'. They describe a period when "the 'abnormality' of the melancholic could consist in abnormal talent" (p. 31), related to an image of "the outstanding man hurled back and forth between exaltation and overwhelming depression" (p. 42). In their book, they describe the slow erosion of anything positive associated with melancholy:

> in future times it was to mean, unambiguously, a bad disposition
> in which unpleasant traits of mind and character were combined
> with poor physique [...] During the first twelve hundred years after
> Christ the idea of the highly gifted melancholic had apparently
> been completely forgotten. (p. 67)

In the Dürer image, Klibansky *et al.* see a vindication of an 'inspired intellectual melancholy' strand that their extensive work had uncovered. Melancholy is personified here, in an image crossed with *ars geometrica,* one of the seven liberal arts. And they further claim "an inner affinity between the two themes" (p. 332). The general tale they have to tell is a portrayal of a realisation of failed effort, of lack of achievement, of someone:

> [whose] mind is preoccupied with interior visions [...] Melencolia's
> eyes stare into the realm of the invisible with the same vain inten
> sity as that with which her hand grasps the impalpable. With [...]
> her gaze thoughtful and sad, fixed on a point in the distance, she
> keeps watch, withdrawn from the world. (pp. 318-320)

[25] Ironically, by the twentieth century, mathematicians had claimed for themselves the right to create objects independent of their physical manifestation (a right previously accorded only to God). Lieven Jonckheere (1991), a Lacanian analyst, writes:

> Henry of Ghent makes an important clinical remark: in his experi
> ence, in the experience of his time, this limitation to spatial repre
> sentation or form, this metaphysical incapacity of immediate access
> to matter, is responsible for the melancholic disposition of the
> mathematical mind. He who tries to measure and calculate all of
> the *materia,* will get depressed. We could also say, referring to
> Lacan, that mathematics constitute[s] a cowardice towards matter.
> (p. 1)

[26] It is noteworthy that, in the expanded edition of Le Lionnais's collection *Great Currents of Mathematical Thought,* space was given to a chapter by Jean Ullmo (1962/1971) on Pascal's distinction. (Recall, too, Magritte's painting, originally entitled 'Maternity', was later retitled by him 'The mathematical mind' – see Figure 7 in Chapter 9.) Interestingly, though perhaps unsurprisingly given its historical location and time, Ullmo tries first to disparage what was translated as 'the subtle intelligence' (*l'esprit de finesse*) and to ensure that the other in revised form (which Ullmo terms the 'scientific intelligence') is seen as the only true one, thereby dissolving Pascal's attempted distinction in a particularly one-sided way.

[27] Giambattista Vico had been less sanguine about the possibility of a 'sensible mathematics', declaring in the first half of the eighteenth century, "Mathematics is created in the self-alienation of the human spirit. The spirit cannot discover itself in

mathematics." (in Davis and Hersh, 1986, p. x). Consequently, mathematician Jean D'Alembert, who worked with Diderot on their Encyclpaedia and wrote its entry for 'blind', needed to defend "mathematicians against the popular idea that mathematics and sensibility were mutually exclusive" (Riskin, 2002, p. 58).

To use a medieval scholastic term, for us there remains the firm possibility of creating a *circumincession*:

> a "mutual indwelling," not of form and content but rather of feeling and thought. […] It may be wrong to think of Cowley, Donne and Chapman as philosopher-poets, but they are poets who knew the mind is a sensory organ like the eye, and for whom there was no aesthetic experience like thinking. (Bringhurst, 2002, pp. 83, 87)

This affinity, between feeling and thought, between and among the various human senses, attuned both to without and within, to the structure of the world and, *pace* Vico, the psychic structure of the self, potentially links the twin sources of mathematics, allowing it to regain and retain an unashamedly aesthetic core.

[28] Autism specialist Uta Frith (2003, p. 25), who first translated Asperger's seminal paper into English, also identifies Erdös as a likely individual with autism, basing her cautious suggestion on biographical information contained in Paul Hoffman's (1998) curiously titled book, *The Man Who Loved Only Numbers*.

[29] While there is some resonance with Worringer's similarly-named categories – for instance, Baron-Cohen's 'systemizing' being closely related to detachment and abstraction – there is also difference: the empathy at play here is specifically empathy with other humans rather than nature and such empathy requires a degree of attachment.

[30] The term 'mind' has just one entry in the index of this book:

> *Changeux:* They [philosophers and psychologists of the "functionalist" school] distinguish the neural organization of the brain from what Anglo-Saxon authors call "mind" (the English term carries none of the metaphysical connotation associated with the French term *esprit*)—that is to say, they distinguish the brain's neural organization from its *functions*. (p. 82; *italics in original*)

[31] David Bodanis (1988), in his essay 'Socialism, bacteria and the obsessions of Pasteur', discusses an similar fear in Louis Pasteur, Frege's somewhat older contemporary. This signal unease in Pasteur's case concerned contamination of the 'body politic' by one agent just as that of the body could be achieved by means of pollution by effectively 'immortal' bacteria. Frege seemed more worried about the 'mind mathematic' being invaded by teeming individual and idiosyncratic psyches, thereby thwarting his desire for the unchanging immortality of mathematical truth.

David Bloor (1976) claimed that Frege's polemical concern with 'psychologism' in the foundations of arithmetic is "steeped in the rhetoric of purity and danger" (p. 83). For more discussion of these notions of Mary Douglas, see Chapter 8 in our book. Bloor immediately went on to summarise Douglas's (1970) 'purity rule', which links high status and strong social control to rigid bodily control:

> The attempt is made to portray interactions as if they are between disembodied spirits. Style and behaviour are bent towards maximising the distance between an activity and its physiological origin. (p. 83)

Chapter 6 documented mathematician Saunders Mac Lane's (1986) similarly 'psychic' (but unpathologised) account of the human origins of mathematical concepts and processes. Nevertheless, this earlier dispute serves as a reminder of the lengths some will go to in order, in T. S. Eliot's (1944, p. 39) words, 'to purify the dialect of the tribe'.

In Greek mythology, Psyche herself was originally human and her tale neatly intertwines not only elements of beauty and pleasure by means of Aphrodite and her son Eros, but also that of a human achieving the divine. In terms of the particular psychology of *mathematics*, it is perhaps also about the investment of human wishes for immortality. As mathematician Marcus du Sautoy (2005) has claimed (albeit without any exemplification, personal or otherwise), "The permanence of mathematical proof fuels the mathematician's belief that, of all the scientists, they alone can achieve immortality" (p. 16).

[32] However, at the very end of his manuscript entitled *The Origin of Geometry*, Edmund Husserl (1936/1970) cautioned:

> For romantic spirits the mythical-magical elements of the historical and prehistorical aspects of mathematics may be particularly attractive; but to cling to this merely historically factual aspect of mathematics is precisely to lose oneself to a sort of romanticism and to overlook the genuine problem, the internal-historical problem, the epistemological problem. (p. 378)

[33] For a range of sources on ethnomathematics, see the 1994 special issue **14**(2) of the journal *For the Learning of Mathematics,* edited by Ubiratan D'Ambrosio and Marcia Ascher, or the collection entitled *Ethnomathematics: Challenging Eurocentrism in Mathematics Education,* edited by Arthur Powell and Marilyn Frankenstein (1997).

REFERENCES

Albers, D. and Alexanderson, G. (eds) (1985) *Mathematical People: Profiles and Interviews,* Boston, MA, Birkhäuser.

Albers, D., Alexanderson, G. and Reid, C. (eds) (1990) *More Mathematical People: Contemporary Conversations,* Boston, MA, Harcourt Brace Jovanovich.

Amiel, H.-F. (1885) *Amiel's Journal: the* Journal Intime *of Henri-Frédéric Amiel* (Ward, M., trans.), London, Macmillan.

Antin, D. (1987) 'The stranger at the door', *Genre* **20**(3/4), 463-481.

Apostol, T. (2000) 'Irrationality of the square root of two: a geometric proof', *The American Mathematical Monthly* **107**(9), 241-242.

Arber, A. (1954) *The Mind and the Eye: a Study of the Biologist's Standpoint,* Cambridge, Cambridge University Press.

Aristotle *Metaphysics* (Apostle, H., trans.), Bloomington, IN, Indiana University Press, 1966.

Asperger, H. (1944/1991) '"Autistic psychopathy" in childhood', in Frith, U. (ed.), *Autism and Asperger Syndrome,* Cambridge, Cambridge University Press, pp. 37-92.

Aston, M. (1988) *England's Iconoclasts: Volume 1 Laws against Images,* Oxford, Clarendon Press.

Aubin, D. (1997) 'The withering immortality of Nicolas Bourbaki: a cultural connector at the confluence of mathematics, structuralism and the Oulipo in France', *Science in Context* **10**(2), 297-342.

Auburn, D. (2001) *Proof,* London, Faber and Faber.

Austin, J. (1962) *How to Do Things with Words,* Cambridge, MA, Harvard University Press.

Ayer, A. (1952) *Language, Truth and Logic,* New York, NY, Dover.

Bailey, D., Borwein, P. and Plouffe, S. (1997) 'On the rapid computation of various polylogarithmic constants', *Mathematics of Computation* **66**(218), 903-913.

Balacheff, N. (1988) 'Aspects of proof in pupils' practice of school mathematics' (Pimm, D., trans.), in Pimm, D. (ed.), *Mathematics, Teachers and Children,* London, Hodder and Stoughton, pp. 216-235.

Balint, M. (1968) *The Basic Fault: Therapeutic Aspects of Regression,* London, Tavistock Publications.

Barabasi, A. (2002) *Linked: the New Science of Networks,* Cambridge, MA, Perseus.

Baron-Cohen, S. (2003) *The Essential Difference: Men, Women and the Extreme Male Brain,* Harmondsworth, Middx, Penguin Books.

Baron-Cohen, S. and Wheelwright, S. (2004) 'The empathy quotient: an investigation of adults with Asperger Syndrome or High Functioning Autism and normal sex differences', *Journal of Autism and Developmental Disorders* **34**(2), 163-175.

Barrow, J. (1995) *The Artful Universe,* Oxford, Oxford University Press.

Barthes, R. (1975) *The Pleasure of the Text,* New York, NY, Noonday Press.

Bartolini Bussi, M. and Boni, M. (2003) 'Instruments for semiotic mediation in primary school classrooms', *For the Learning of Mathematics* **23**(2), 15-22.

Bateson, G. (1979) *Mind and Nature: a Necessary Unity,* New York, NY, Dutton.

Bateson, G. (1991) A *Sacred Unity: Further Steps to an Ecology of Mind,* New York, NY, HarperCollins.

Baudelaire, C. (1861/1995) *Les Fleurs du Mal,* Paris, Slatkine.

Baumgarten, A. (1750/1961) *Aesthetica,* Hildesheim, G. Olms.

Baxandall, M. (1985) *Patterns of Intention: on the Historical Explanation of Pictures,* New Haven, CT, Yale University Press.

Beardsley, M. (1982) *The Aesthetic Point of View,* Ithaca, NY, Cornell University Press.

Beaulieu, L. (2000) 'Bourbaki's art of memory', in Abir-Am, P. and Elliott, C. (eds), *Commemorative Practices in Science: Historical Perspectives on the Politics of Collective Memory* (*Osiris* v. 14), Ithaca, NY, Cornell University, pp. 219-251.

Becker, O. (1957) *Zwei Untersuchungen zur antiken Logik,* Wiesbaden, Harassowitz. [Two Enquiries into Ancient Logic]

Bell, C. (1914/1992) 'The aesthetic hypothesis', in Harrison, C. and Wood, P. (eds), *Art In Theory 1900-1990: an Anthology of Changing Ideas,* Oxford, Blackwell, pp. 113-116.

Beltrami, E. (1999) *What is Order? Chance and Order in Mathematics and Life,* New York, NY, Copernicus.

Berger, J. (1972) *Ways of Seeing,* Harmondsworth, Middx, Penguin.

Berlinski, D. (1995) *A Tour of the Calculus,* New York, NY, Vintage.

Berlinski, D. (1997) 'Ground zero' (a review of *The Pleasures of Counting* by T. W. Körner), *The Sciences* **37**(4), 37-41.

Bernard, E. (1926) *Souvenirs sur Paul Cézanne: une Conversation avec Cézanne,* Paris, R. G. Michel.

Besançon, A. (2000) *The Forbidden Image: an Intellectual History of Iconoclasm* (Todd, J., trans.), Chicago, IL, University of Chicago Press.

Bill, M. (1949/1993) 'The mathematical way of thinking in the visual art of our time', in Emmer, M. (ed.), *The Visual Mind: Art and Mathematics,* Boston, MA, The MIT Press, pp. 5-9.

Birkhoff, G. (1933) *Aesthetic Measure,* Cambridge, MA, Harvard University Press.

Birkhoff, G. (1933/1956) 'Mathematics of aesthetics', in Newman, J. (ed.), *The World of Mathematics,* New York, NY, Simon and Schuster, vol. 4, pp. 2185-2195.

Bishop, A. (1991) *Mathematical Enculturation: a Cultural Perspective on Mathematics Education,* Dordrecht, Kluwer Academic Publishers.

Blake, W. (1810/1967) 'A vision of the last judgment' (extracts), in Perkins, D. (ed.) *English Romantic Writers,* New York, NY, Harcourt, Brace and World, pp. 161-163.

Bloor, D. (1976) *Knowledge and Social Imagery,* London, Routledge and Kegan Paul.

Bodanis, D. (1988) *Web of Words: the Ideas behind Politics,* Basingstoke, Hants, Macmillan.

Boer, C. (1980) 'Introduction', in *Marsilio Ficino: the Book of Life* (Boer, C., trans.), Irving, TX, Spring Publications, pp. iii-xix.

Bollas, C. (1987) *The Shadow of the Object: Psychoanalysis of the Unthought Known*, London, Free Association Books.

Bollas, C. (1999) *The Mystery of Things*, London, Routledge.

Bollas, C. (2002) *Free Associations*, Cambridge, Icon Books.

Borel, A. (1983) 'Mathematics: art and science', *The Mathematical Intelligencer* **5**(4), 9-17.

Borwein, J. and Bailey, D. (2003) *Mathematics by Experiment: Plausible Reasoning in the 21st Century*, Natick, MA, A. K. Peters.

Borwein, J. and Borwein, P. (1984) 'The arithmetic–geometric mean and fast computation of elementary functions', *SIAM Review* **26**(2), 351-366.

Borwein, J. and Borwein, P. (2001) 'Challenges for mathematical computing', *Computing in Science and Engineering* **3**(3), 48-53.

Borwein, J. and Corless, R. (1999) 'Emerging tools for experimental mathematics', *The American Mathematical Monthly* **106**(10), 889-909.

Borwein, J. and Lisonek, P. (2000) 'Applications of integer relation algorithms', *Discrete Mathematics* **217**(1), 65-82.

Bouleau, N. (1997) *Dialogues autour de la Création mathématique*, Paris, Association Laplace-Gauss.

Bouleau, N. (2002) *La Règle, le Compas et le Divan: Plaisirs et Passions mathématiques*, Paris, Editions du Seuil.

Bourbaki, N. (1960a, 2nd edn) *Théorie des Ensembles* (Fasicule XVII), Paris, Hermann.

Bourbaki, N. (1960b) *Éléments d'Histoire des Mathématiques*, Paris, Hermann.

Boyer, C. (1959) *The Rainbow: from Myth to Mathematics*, New York, NY, Thomas Yoseloff.

Brent, J. (1998, rev'd edn) *Charles Sanders Peirce: a Life*, Bloomington, IN, Indiana University Press.

Bringhurst, R. (2002) 'The philosophy of poetry and the trashing of Doctor Empedokles', in Lilburn, T. (ed.), *Thinking and Singing: Poetry and the Practice of Philosophy*, Toronto, ON, Cormorant Books, pp. 79-93.

Brookes, W. (1970) 'Preface', in *Mathematical Reflections*, Cambridge, Cambridge University Press, p. vii.

Brouwer, L. (1908/1975) 'The unreliability of the logical principles', in *Collected Works* (Heyting, A., ed.), vol. 1, New York, NY, American Elsevier, pp. 107-111.

Brown, J. (1999) *Philosophy of Mathematics: an Introduction to the World of Proofs and Pictures*, New York, NY, Routledge.

Bullough, E. (1963) '"Psychical Distance" as a factor in art and aesthetic principle', in Levich, M. (ed.), *Aesthetics and the Philosophy of Criticism*, New York, NY, Random House, pp. 233-254.

Burkert, W. (1972) *Lore and Science in Ancient Pythagoreanism* (Minar, E., trans.), Cambridge, MA, Harvard University Press.

Burton, L. (2004) *Mathematicians as Enquirers: Learning about Learning Mathematics*, Dordrecht, Kluwer Academic Publishers.

Butterworth, B. (1999) *What Counts: How Every Brain Is Hardwired for Math*, New York, NY, Free Press.

Campbell, P. (1989) 'The geometry of decoration on prehistoric Pueblo pottery from Starkweather Ruin' (Symmetry 2: unifying human understanding, Part 2), *Computer Mathematics and Its Applications* **17**(4-6), 731-749.

Carlson, A. (2000) *Aesthetics and the Environment: the Appreciation of Nature, Art and Architecture,* New York, NY, Routledge.

Chaitin, G. (2002) www.cs.auckland.ac.nz/CDMTCS/chaitin/coimbra.pdf.

Chandler, B. and Magnus, W. (1986) 'Comments on an essay entitled "The modern algebraic method" by H. Hasse published in 1929', *The Mathematical Intelligencer* **8**(2), 23-25.

Chandrasekhar, S. (1987) *Truth and Beauty: Aesthetics and Motivations in Science,* Chicago, IL, University of Chicago Press.

Changeux, J.-P. and Connes, A. (1995) *Conversations on Mind, Matter, and Mathematics,* Princeton, NJ, Princeton University Press.

Charraud, N. (1994) *Infini et Inconscient: Essai sur Georg Cantor,* Paris, Anthropos.

Charraud, N. (1997) *Lacan et les Mathématiques,* Paris, Anthropos.

Chevallard, Y. (1990) 'On mathematics education and culture: critical afterthoughts', *Educational Studies in Mathematics* **21**(1), 3-27.

Chorbachi, W. (1989) 'In the tower of Babel: beyond symmetry in Islamic design' (Symmetry 2: unifying human understanding, Part 2), *Computer Mathematics and Its Applications* **17**(4–6), 751-789.

Chouchan, M. (1995) *Nicolas Bourbaki: Faits et Légendes,* Argenteuil, Editions du Choix.

Clark, K. (1951) *Piero della Francesca,* London, Phaidon.

Cole, K. (1998) *The Universe and the Teacup: the Mathematics of Truth and Beauty,* New York, NY, Harcourt Brace.

Comte, F. (1994) *The Wordsworth Dictionary of Mythology,* Ware, Herts, Wordsworth Editions, Ltd.

Corry, L. (1996) *Modern Algebra and the Rise of Mathematical Structures,* Basel, Birkhäuser.

Corry, L. (1997) 'The origins of eternal truth in modern mathematics: Hilbert to Bourbaki and beyond', *Science in Context* **10**(2), 253-296.

Coxeter, H. (1948) 'A problem of collinear points', *The American Mathematical Monthly* **55**(1), 26-28.

Cran, B. (2002) *The Good Life,* Roberts Creek, BC, Nightwood Editions.

Critchlow, K. (1979) *Time Stands Still: New Light on Megalithic Science,* London, Gordon Fraser.

Cromwell, P. (1997) *Polyhedra,* Cambridge, Cambridge University Press.

Curtin, D. (ed.) (1982) *The Aesthetic Dimension of Science,* New York, NY, Philosophical Library.

Damasio, A. (1994) *Descartes' Error: Emotion, Reason, and the Human Brain,* New York, NY, Avon Books.

Darwin, C. (1887/1958) *The Autobiography of Charles Darwin and Selected Letters* (Darwin, F., ed.), New York, NY, Dover.

Davis, P. (1997) *Mathematical Encounters of the Second Kind,* Boston, MA, Birkhäuser.

Davis, P. and Hersh, R. (1980) *The Mathematical Experience,* Boston, MA, Birkhäuser.

Davis, P. and Hersh, R. (1986) *Descartes' Dream: the World According to Mathematics,* Boston, MA, Houghton-Mifflin.

Dawkins, R. (1986) *The Blind Watchmaker: Why Evidence of Evolution Reveals a Universe Without Design,* New York, NY, Norton.

de Villiers, M. (1999) *Rethinking Proof with the Geometer's Sketchpad,* Berkeley, CA, Key Curriculum Press.

Dedekind, R. (1872/1963) *Essays on the Theory of Numbers,* New York, NY, Dover.

Degenaar, M. (1996) *Molyneux's Problem: Three Centuries of Discussion on the Perception of Forms,* Dordrecht, Kluwer Academic Publishers.

Dehaene, S. (1997) *The Number Sense: How the Mind Creates Mathematics,* Oxford, Oxford University Press.

Dehaene, P. *et al.* (1999) 'Sources of mathematical thinking: behavioral and brain-imaging evidence', *Science* **284**(5416), 970-974.

Dennis, D. and Confrey, J. (1997) 'Drawing logarithmic curves with *Geometer's Sketchpad*: a method inspired by historical sources', in Schattschneider, D. and King, J. (eds), *Geometry Turned On: Dynamic Software in Learning, Teaching, and Research,* Washington, DC, The Mathematical Association of America, pp. 147-156.

Derbyshire, J. (2003) *Prime Obsession: Bernhard Riemann and the Greatest Unsolved Problem in Mathematics,* Washington, DC, Joseph Henry Press.

Derrida, J. (1978/1987) *The Truth in Painting* (Bennington, G. and McLeod, I., trans.), Chicago, IL, University of Chicago Press.

Devlin, K. (1994) *Mathematics: the Science of Patterns (The Search for Order in Life, Mind, and the Universe),* New York, NY, Scientific American Library.

Devlin, K. (2000) *The Math Gene: How Mathematical Thinking Evolved and Why Numbers Are Like Gossip,* New York, NY, Basic Books.

Dewey, J. (1934) *Art as Experience,* New York, NY, Perigree.

Dickson, S. (1993) 'True 3D computer modeling: sculpture of numerical abstraction', in Emmer, M. (ed.), *The Visual Mind: Art and Mathematics,* Cambridge, MA, MIT Press, pp. 93-99.

Diderot, D. (1769/1966) *Rameau's Nephew and D'Alembert's Dream* (Tancock, L., trans.), Harmondsworth, Middx, Penguin.

Dieudonné, J. (1962/1971) 'Modern axiomatic methods and the foundations of mathematics', in Le Lionnais, F. (ed.), *Great Currents of Mathematical Thought,* New York, NY, Dover, pp. 251-266.

Dieudonné, J. (1969) *Linear Algebra and Geometry,* Paris, Hermann.

Dissanayake, E. (1988) *What Is Art for?,* Seattle, WA, University of Washington Press.

Dissanayake, E. (1995) *Homo Aestheticus: Where Art Comes from and Why,* Seattle, WA, University of Washington Press.

Dissanayake, E. (2000) *Art and Intimacy: How the Arts Began,* Seattle, WA, University of Washington Press.

Dixon, R. (1995) *The Baumgarten Corruption: from Sense to Nonsense in Art and Philosophy,* East Haven, CT, Pluto Press.

Dongarra, J. and Sullivan, F. (2000) 'The top 10 algorithms', *Computing in Science and Engineering* **2**(1), 22-23.

Donoghue, D. (2003) *Speaking of Beauty,* New Haven, CT, Yale University Press.

Douglas, M. (1966) *Purity and Danger: an Analysis of Concepts of Pollution and Taboo,* London, Routledge and Kegan Paul.

Douglas, M. (1970) *Natural Symbols: Explorations in Cosmology,* London, Barrie and Rockliff the Cresset.

Doxiadis, A. (2000) *Uncle Petros and Goldbach's Conjecture,* New York, NY, Bloomsbury Publishing.

Drake, S. (1970) *Galileo Studies: Personality, Tradition, and Revolution,* Ann Arbor, MI, University of Michigan Press.

Dreyfus, T. and Eisenberg, T. (1986) 'On the aesthetics of mathematical thought', *For the Learning of Mathematics* **6**(1), 2-10.

du Sautoy, M. (2003) *The Music of the Primes: Searching to Solve the Greatest Mystery in Mathematics,* New York, NY, HarperCollins.

du Sautoy, M. (2005) 'Escape by numbers', *The Guardian,* 29th March, p. 16.

Dummett, M. (1964) 'Wittgenstein's philosophy of mathematics', in Benacerraf, P. and Putnam, H. (eds), *Philosophy of Mathematics: Selected Readings,* Oxford, Blackwell, pp. 491-509.

Dunnington, G. (1955/2004) *Gauss, Titan of Science,* Washington, DC, The Mathematical Association of America.

Dürrenmatt, F. (1989) *Durcheinandertal,* Zürich, Diogenes.

Dyson, F. (1982) 'Manchester and Athens', in Curtin, D. (ed.), *The Aesthetic Dimension of Science,* New York, NY, Philosophical Library, pp. 41-62.

Dyson, F. *et al.* (1982) 'Discussions', in Curtin, D. (ed.), *The Aesthetic Dimension of Science,* New York, NY, Philosophical Library, pp. 107-145.

Efimov, N. (1964) 'Generation of singularities on surfaces of negative curvature', *Matematicheskii Sbornik* (New Series) **64**(106), 286-320. [In Russian]

Ehrenzweig, A. (1967) *The Hidden Order of Art: a Study in the Psychology of Artistic Imagination,* Berkeley, CA, University of California Press.

Eliot, G. (1859/1961) *Adam Bede,* Harmondsworth, Middx, Penguin.

Eliot, T. (1921/1932) 'The metaphysical poets', in *Selected Essays: 1917-32,* New York, NY, Harcourt, Brace and Company, pp. 241-250.

Eliot, T. (1944) 'Little Gidding', in *Four Quartets,* London, Faber and Faber, pp. 35-44.

Emmer, M. and Manaresi, M. (eds) (2003) *Mathematics, Art, Technology and Cinema,* Berlin, Springer-Verlag.

Endress, G. (2003) 'Mathematics and philosophy in the system and practice of science in Islam', in Hogendijk, J. and Sabra, A. (eds), *The Enterprise of Science in Islam: New Perspectives,* Cambridge, MA, MIT Press, pp. 121-176.

Everdell, W. (1997) *The First Moderns: Profiles in the Origins of Twentieth-Century Thought,* Chicago, IL, University of Chicago Press.

Eves, H. (1972, rev'd edn) *A Survey of Geometry,* Boston, MA, Allyn and Bacon.

Eves, H. (1980) 'The thinker and the thug', in *Great Moments in Mathematics (before 1650),* Washington, DC, The Mathematical Association of America, pp. 83-95.

Fadiman, C. (ed.) (1985) *The Little, Brown Book of Anecdotes,* Boston, MA, Little, Brown and Company.

Farmelo, G. (ed.) (2002) *It Must Be Beautiful: Great Equations of Modern Science,* London, Granta Books.

Fauvel, J. (1987) 'Mathematics in the Greek World' (Unit 2), MA290 *Topics in the History of Mathematics,* Milton Keynes, Bucks, The Open University.

Fauvel, J. (1988) 'Cartesian and Euclidean rhetoric', *For the Learning of Mathematics* **8**(1), 25-29.

Fauvel, J. and Gray, J. (eds) (1987) *The History of Mathematics: a Reader,* Basingstoke, Hants, Macmillan.

Featherstone, H. (2000) '"–Pat + Pat = 0": intellectual play in elementary mathematics', *For the Learning of Mathematics* **20**(2), 14-23.

Fénélon, F. de (1697/1845) *Oeuvres de Fénélon,* vol. 1, Paris, Firmin-Didot frères, fils et cie.

Ferguson, E. (1962) 'Kinematics of mechanisms from the time of Watt', *United States National Museum Bulletin* **228,** 185-230.

Ferguson, E. (1992) *Engineering and the Mind's Eye,* Cambridge, MA, MIT Press.

Fischer, E. (1999) *Beauty and the Beast: the Aesthetic Moment in Science,* New York, NY, Plenum.

Fisher, P. (1998) *Wonder, the Rainbow, and the Aesthetics of Rare Experiences,* Cambridge, MA, Harvard University Press.

Fitzgerald, M. (2000) 'Is the cognitive style of the persons with the Asperger's Syndrome also a "mathematical style"?', *Journal of Autism and Developmental Disorders* **30**(2), 175-176.

Fitzgerald, M. (2002) 'Asperger's Disorder and mathematicians of genius', *Journal of Autism and Developmental Disorders* **32**(1), 59-60.

Fitzgerald, M. (2004) *Autism and Creativity: Is there a Link between Autism in Men and Exceptional Ability?,* London, Routledge.

Flannery, S. with Flannery, D. (2001) *In Code: a Mathematical Journey,* New York, NY, Workman.

Fodor, J. (1985) 'Précis of The Modularity of Mind', *The Behavioral and Brain Sciences* **8**(1), 1-42.

Fowler, D. (1985a) '400 years of decimal fractions', *Mathematics Teaching* **110,** 20-21.

Fowler, D. (1985b) '400.25 years of decimal fractions', *Mathematics Teaching* **111,** 30-31.

Fowler, D. (1993) 'The story of the discovery of incommensurability, revisited', in Gavroglu, K., Christianidis, J. and Nicolaidis, E. (eds), *Trends in the Historiography of Science* (Boston Studies in the Philosophy of Science no. 151), Boston, MA, Kluwer, pp. 221-235.

Fowler, D. (1999, 2nd edn) *The Mathematics of Plato's Academy: a New Reconstruction,* Oxford, Clarendon Press.

Fox Talbot, W. (1842/1969) *The Pencil of Nature,* New York, NY, Da Capo Press.

Frayn, M. (1974) *Constructions,* London, Wildwood House.

Frayn, M. (1998) *Copenhagen,* London, Methuen.

Frege, G. (1884/1978; 2nd revd edn) *The Foundations of Arithmetic: a Logico-Mathematical Enquiry into the Concept of Number* (Austin, J., trans.), Oxford, Basil Blackwell.

Freud, S. (1917/1955) 'Mourning and melancholia', in Strachey, J. (ed.), *The Standard Edition of the Complete Psychological Works of Sigmund Freud,* vol. 14, London, Hogarth Press, pp. 243-258.

Freud, S. (1920/1955) 'Beyond the pleasure principle', in Strachey, J. (ed.), *The Standard Edition of the Complete Psychological Works of Sigmund Freud,* vol. 18, London, Hogarth Press, pp. 7-64.

Frith, U. (2003, 2nd edn) *Autism: Explaining the Enigma,* Oxford, Blackwell.

Galilei, G. (1632/1953) *Dialogue Concerning the Two Chief World Systems–Ptolomaic and Copernican* (Drake, S., trans.), Berkeley, CA, University of California Press.

Gardner, H. (1987) *The Mind's New Science: a History of the Cognitive Revolution,* New York, NY, Basic Books.

Gattegno, C. (1970) *What We Owe Children: the Subordination of Teaching to Learning,* New York, NY, Avon Books.

Gattegno, C. (1998) *The Science of Education Part 2b: The Awareness of Mathematization,* New York, NY, Educational Solutions.

Gauss, C. (1801/1966) *Disquisitiones Arithmeticae* (Clarke, A., trans.), New Haven, CT, Yale University Press.

Gauss, C. (1863) *Werke* (Schering, E., ed.), II, Leipzig-Berlin, Königliche Gesellschaft der Wissenschaften zu Göttingen.

Gavin, B. (1990) *Breaking Free of the Earth: Kazimir Malevich 1878–1935,* London, Channel Four/La Sept/RM Arts. [videotape]

Gazalé, M. (1999) *Gnomon: from Pharaohs to Fractals,* Princeton, NJ, Princeton University Press.

Gelernter, D. (1998) *Machine Beauty: Elegance and the Heart of Technology,* New York, NY, Basic Books.

Gerdes, P. (2003) *Awakening of Geometrical Thought in Early Culture,* Minneapolis, MN, MEP Publications.

Gerofsky, S. (1996) 'A linguistic and narrative view of word problems in mathematics education', *For the Learning of Mathematics* **16**(2), 36-45.

Gerofsky, S. (2004) *A Man Left Albuquerque Heading East: Word Problems as Genre in Mathematics Education,* New York, NY, Peter Lang.

Gibbs, N., Lacayo, R., Morrow, R. and Smolowe, J. (eds) (1996) *Mad Genius: the Odyssey, Pursuit, and Capture of the Unabomber Suspect,* New York, NY, Warner.

Gleick, J. (1987) *Chaos: Making a New Science,* New York, NY, Viking Penguin.

Gleick, J. (2003) *Isaac Newton,* New York, NY, Pantheon Books.

Gombrich, E. (1960) *Art and Illusion: a Study in the Psychology of Pictorial Representation,* New York, NY, Pantheon Books.

Gombrich, E. (1979) *The Sense of Order: a Study in the Psychology of Decorative Art,* Oxford, Phaidon Press.

Goodman, N. (1983) 'Reflections on Bishop's philosophy of mathematics', *The Mathematical Intelligencer* **5**(3), 61-68.

Gowers, T. (2002) *Mathematics: a Very Short Introduction,* Oxford, Oxford University Press.

Graham-Dixon, A. (1996) *A History of British Art,* London, BBC Books.

Grattan-Guinness, I. (1997) *The Fontana History of the Mathematical Sciences: the Rainbow of Mathematics,* London, Fontana Press.

Gray, J. (1984) 'A commentary on Gauss's mathematical diary, 1796–1814, with an English translation', *Expositiones Mathematicae* **2**(2), 97-130.

Greenberg, C. (1939) 'Avant-garde and kitsch', *Partisan Review* **6**(5), 34-49.

Greenberg, C. (1960/1965) 'Modernist painting', *Art and Literature* **4**, 193-201.

Greenberg, C. (1978/1999) 'The experience of value', in *Homemade Esthetics: Observations on Art and Taste,* New York, NY, Oxford University Press, pp. 59-64.

Grünbaum, B. and Shephard, G. (1993) 'Interlace patterns in Islamic and Moorish art', in Emmer, M. (ed.), *The Visual Mind: Art and Mathematics,* Cambridge, MA, The MIT Press, pp. 147-155.

Guedj, D. (1981/1985) 'Nicholas [*sic*] Bourbaki, collective mathematician: an interview with Claude Chevalley' (Gray, J., trans.), *The Mathematical Intelligencer* **7**(2), 18-22.

Gustin, W. (1985) 'The development of exceptional research mathematicians', in Bloom, B. (ed.), *Developing Talent in Young People,* New York, NY, Ballantine, pp. 270-331.

Gutch, J. (ed.) (1796) *The History and Antiquities of the University of Oxford,* in Two Books by Anthony à Wood. M.A. of Merton College (now first published in English from the original MS in the Bodleian Library by John Gutch. M.A. Chaplain of All Souls and Corpus Christi Colleges), vol. 2, part 1, Oxford, printed for the editor.

Guy, R. (1988) 'The strong law of small numbers', *The American Mathematical Monthly* **95**(8), 697-712.

Hadamard, J. (1945) T*he Psychology of Invention in the Mathematical Field,* Princeton, NJ, Princeton University Press.

Haddon, M. (2003) *The Curious Incident of the Dog in the Night-Time,* New York, NY, Doubleday.

Hanson, N. (1958) *Patterns of Discovery: an Inquiry into the Conceptual Foundations of Science,* Cambridge, Cambridge University Press.

Hardy, G. (1940) *A Mathematician's Apology,* Cambridge, Cambridge University Press.

Hardy, G. (1945/1999, 3rd edn) *Ramanujan: Twelve Lectures on Subjects Suggested by His Life and Work,* New York, NY, Chelsea.

Harré, R. (1958) 'Quasi-aesthetic appraisals', *Philosophy* **33**(125), 132-137.

Hasse, H. (1930/1986) 'The modern algebraic method', *The Mathematical Intelligencer* **8**(2), 18-23.

Havil, J. (2003) *Gamma: Exploring Euler's Constant,* Princeton, NJ, Princeton University Press.

Heath, T. (ed.) (1926/1956) *Euclid's Elements,* New York, NY, Dover.

Heilbron, J. (1998) *Geometry Civilized: History, Culture, Technique,* Oxford, Oxford University Press.

Henderson, D. (1973) 'A simplicial complex whose product with any ANR is a simplicial complex', *General Topology and Its Applications* **3**(1), 81-83.

Henderson, D. and Taimina, D. (1998) *Differential Geometry: a Geometric Introduction,* Upper Saddle River, NJ, Prentice Hall.

Henderson, D. and Taimina, D. (2001a) *Experiencing Geometry: in Euclidean, Spherical, and Hyperbolic Spaces,* Upper Saddle River, NJ, Prentice Hall.

Henderson, D. and Taimina, D. (2001b) 'Crocheting the hyperbolic plane', *The Mathematical Intelligencer* **23**(2), 17-28.

Henderson, D. and Taimina, D. (2005a) *Experiencing Geometry: Euclidean and Non-Euclidean with History,* Upper Saddle River, NJ, Prentice-Hall.

Henderson, D. and Taimina, D. (2005b) 'How to use history to clarify common confusions in geometry', in Shell-Gellasch, A. and Jardine, D. (eds), *Using Recent History of Mathematics in Teaching Mathematics: Ideas for Incorporating the Last 200 Years of Mathematics History into the Classroom,* Washington, DC, The Mathematical Association of America, pp. 57-73.

Henwood, M. and Rival, I. (1979) 'Eponymy in mathematical nomenclature: what's in a name, and what should be?', *The Mathematical Intelligencer* **2**(4), 204-205.

Herbst, P. (2002) 'Establishing a custom of proving in American school geometry: evolution of the two-column proof in the early twentieth century', *Educational Studies in Mathematics* **49**(3), 283-312.

Hermite, C. (1893/1905) 'Lettre à Stieltjes, 20 Mai, 1893', in Baillaud, B. and Bourget, H. (eds), *Correspondance d'Hermite et de Stieltjes,* Paris, Gauthier-Villars, vol. 2, pp. 317-319.

Hersh, R. (1995) 'Fresh breezes in the philosophy of mathematics', *The American Mathematical Monthly* **102**(7), 589-594.

Hett, W. (ed. and trans.) (1957) *Problems,* 2 vols, Cambridge, MA, Harvard University Press.

Hewitt, D. (1992) 'Train spotters' paradise', *Mathematics Teaching* **140**, 6-8.

Higgins, P. (1998) *Mathematics for the Curious,* Oxford, Oxford University Press.

Hilbert, D. (1901) 'Ueber Flächen von constanter Gaussscher Krümmung', *Transactions of the American Mathematical Society* **2,** 87-99.

Hilbert, D. and Cohn-Vossen, S. (1932/1983) *Geometry and the Imagination,* New York, NY, Chelsea Publishing Co.

Hildebrand, G. (1999) *Origins of Architectural Pleasure,* Berkeley, CA, University of California Press.

Hoban, R. (1982) *Riddley Walker,* London, Picador.

Hoffman, P. (1999) *The Man Who Loved Only Numbers: the Story of Paul Erdös and the Search for Mathematical Truth,* Boston, MA, Little, Brown.

Hofstadter, D. (1992) 'From Euler to Ulam: discovery and dissection of a geometric gem' (pre-print version available from the author). An abbreviated version of this paper was subsequently published under the following title 'Discovery and dissection of a geometric gem' – see next reference.)

Hofstadter, D. (1997) 'Discovery and dissection of a geometric gem', in Schattschneider, D. and King, J. (eds), *Geometry Turned On: Dynamic Software in Learning, Teaching, and Research,* Washington, DC, The Mathematical Association of America, pp. 3-14.

Hogben, L. (1940) *Mathematics for the Million,* New York, NY, Norton.

Honsberger, R. (1973) *Mathematical Gems,* Washington, DC, The Mathematical Association of America.

Hoodbhoy, P. (1991) *Islam and Science: Religious Orthodoxy and the Battle for Rationality,* London, Zed Books.

Howson, G. (1984) 'Communications', *For the Learning of Mathematics* **4**(1), 41-42.

Huber-Dyson, V. (1998) 'On the nature of mathematical concepts: why and how do mathematicians jump to conclusions?', *Edge* **34,** February 16. (Available from www.edge.org/3rd_culture/huberdyson/.)

Hudson, L. (1970) *Frames of Mind: Ability, Perception and Self-Perception in the Arts and Sciences,* Harmondsworth, Middx, Penguin.

Huffman, C. (1993) *Philolaus of Croton: Pythagorean and Presocratic,* Cambridge, Cambridge University Press.

Hughes-Hallet, P. (2000) *The Immortal Dinner: a Famous Dinner of Genius and Laughter in Literary London, 1817,* London, Viking.

Huizinga, J. (1950) *Homo Ludens: a Study of the Play Element in Culture,* London, Routledge and Kegan Paul.

Husserl, E. (1936/1970) *The Crisis of European Sciences and Transcendental Phenomenology* (Carr, D., trans.), Evanston, IL, Northwestern University Press.

Idzerda, S. (1954) 'Iconoclasm during the French Revolution', *American Historical Review* **60**(1), 13-26.

Illich, I. (1994) *The Wisdom of Leopold Kohr,* Fourteenth Annual E. F. Schumacher Lectures, New Haven, CT, Yale University. (Available from: www.smallisbeautiful.org/lec-ill.html.)

Ivins, W. (1946) *Art and Geometry: a Study in Space Intuitions,* Cambridge, MA, Harvard University Press.

Ivins, W. (1969) *Prints and Visual Communication,* New York, NY, Da Capo Press.

Jackiw, N. (1991, 2001) *The Geometer's Sketchpad,* Emeryville, CA, Key Curriculum Press. (software)

Jacobs, J. (1984) *Cities and the Wealth of Nations: Principles of Economic Life,* New York, NY, Random House.

James, I. (2002) *Remarkable Mathematicians: from Euler to von Neumann,* Cambridge, The Mathematical Association of America/Cambridge University Press.

James, I. (2003) 'Autism in mathematicians', *The Mathematical Intelligencer* **25**(4), 62-65.

James, I. (2004) *Remarkable Physicists: from Galileo to Yukawa,* Cambridge, Cambridge University Press.

James, W. (1902) *The Varieties of Religious Experience: a Study in Human Nature,* New York, NY, Modern Library.

Jonckheere, L. (1991) www.lacanian.net/Ornicar%20online/Archive%20OD/ornicar/articles/lj0206.htm.

Jones, D. (1952) *The Anathemata: Fragments of an Attempted Writing,* London, Faber and Faber.

Josipovici, G. (1996) *Touch,* New Haven, CT, Yale University Press.

Joyce, J. (1939/1959) *Finnegan's Wake,* New York, NY, Viking.

Judson, H. (1979) *The Eighth Day of Creation: the Makers of the Revolution in Biology,* New York, NY, Simon and Schuster.

Judson, H. (1980a) *The Search for Solutions,* New York, NY, Holt, Rinehart and Winston.

Judson, H. (1980b) 'The rage to know: science as the true modern art', *The Atlantic Monthly,* **245**(4), 112-117.

Kandinsky, W. (1926/1979) *Point and Line to Plane,* New York, NY, Dover.

Kandinsky, W. (1931) 'Réflexions sur l'art abstrait', *Cahiers d'Art* **7/8**, 351-353.

Kant, I. (1790/1987) *Critique of Judgment* (Pluhar, W., trans.), Indianapolis, IN, Hackett Publishing.

Kaplan, R. (2001) *The Nothing That Is: a Natural History of Zero,* New York, NY, Oxford University Press.

Kemp, M. (2000) *Visualizations: the Nature Book of Art and Science,* Oxford, Oxford University Press.

Keys, J. (1971) *Only Two Can Play This Game,* Cambridge, Cat Books.

King, J. (1992) *The Art of Mathematics,* New York, NY, Plenum Press.

Klein, J. (1968) *Greek Mathematical Thought and the Origin of Algebra,* Cambridge, MA, MIT Press.

Klein, J. (1987) *Our Need for Others and Its Roots in Infancy,* London, Tavistock Publications.

Klibansky, R., Panofsky, E. and Saxl, F. (1964) *Saturn and Melancholy: Studies in the History of Natural Philosophy, Religion and Art,* London, Nelson.

Kline, M. (1953) *Mathematics in Western Culture,* London, George Allen and Unwin.

Kline, M. (1972) *Mathematical Thought from Ancient to Modern Times,* New York, NY, Oxford University Press.

Knorr, W. (1975) *The Evolution of the Euclidean Elements: a Study of the Theory of Incommensurable Magnitudes and Its Significance for Early Greek Geometry,* Dordrecht, D. Reidel.

Krull, W. (1930/1987) 'The aesthetic viewpoint in mathematics', *The Mathematical Intelligencer* **9**(1), 48-52.

Kuiper, N. (1955) 'On C¹-isometric embeddings, ii', *Nederlandse Akademie van Wetenschappen* (Proc. Ser. A) **58**, 683-689.

La Mettrie, J. de (1748/1912) *Man a Machine* (Bussey, G., trans.), Chicago, IL, Open Court.

Laborde, J.-M., Baulac, Y. and Bellemain, F. (1989) *Cabri-Géomètre,* Grenoble, LSD2-IMAG. (software)

Lachterman, D. (1989) *The Ethics of Geometry: a Genealogy of Modernity,* New York, NY, Routledge.

Lakatos, I. (1976) *Proofs and Refutations: the Logic of Mathematical Discovery,* Cambridge, Cambridge University Press.

Lakoff, G. and Johnson, M. (1999) *Philosophy in the Flesh: the Embodied Mind and Its Challenge to Western Thought,* New York, NY, Basic Books.

Lakoff, G. and Nuñez, R. (2000) *Where Mathematics Comes from: How the Embodied Mind Brings Mathematics into Being,* New York, NY, Basic Books.

Lane, P. (1982) 'A red bird bearing on his back an empty cup', in *Old Mother,* Toronto, ON, Oxford University Press, pp. 44-45.

Latour, B. (2002) 'What is iconoclash? Or is there a world beyond the image wars?', in Latour, B. and Weibel, P. (eds), *Iconoclash: Beyond the Image Wars in Science, Religion, and Art,* Cambridge, MA, MIT Press, pp. 14-37.

Le Lionnais, F. (ed.) (1948) *Les grands Courants de la Pensée mathématique,* Marseille, Cahiers du Sud.

Le Lionnais, F. (1948/1971) 'Beauty in mathematics', in Le Lionnais, F. (ed.), *Great Currents of Mathematical Thought,* New York, NY, Dover, pp. 121-158.

Le Lionnais, F. (1948/1986) 'La beauté en mathématiques', in Le Lionnais, F. (ed.), *Les grands Courants de la Pensée mathématique Paris,* Editions Rivages, pp. 437-465.

Le Lionnais, F. (ed.) (1962, 2nd edn) *Les grands Courants de la Pensée mathématique,* Paris, Librairie Scientifique et Technique.

Le Lionnais, F. with Brette, J. (1983) *Les Nombres Remarquables,* Paris, Hermann.

Lebesgue, H. (1941) 'Les controverses sur la théorie des ensembles et la question des fondements', in Gonseth, F. (ed.) *Les Entretiens de Zurich,* Zurich, S. A. Leeman, pp. 109-122.

Levinson, J. (ed.) (1998) *Aesthetics and Ethics,* Cambridge, Cambridge University Press.

Levinson, S. (1983) *Pragmatics,* Cambridge, Cambridge University Press.

Lévy, P. (1970) *Quelques Aspects de la Pensée d'un Mathématicien,* Paris, Albert Blanchard.

Lewis, C. (1964) *The Discarded Image: an Introduction to Medieval and Renaissance Literature,* Cambridge, Cambridge University Press.

Lewis-Williams, D. (2002) *The Mind in the Cave,* London, Thames and Hudson.

Lewontin, R. (2001) 'In the beginning was the word', *Science* **291**(5507), 1263-1264.

Lissitsky, E. (1925/1968) 'A. and pangeometry', in Lissitisky-Küppers, S. (ed.), *Life, Letters, Texts,* Greenwich, CT, New York Graphic Society, pp. 348-354.

Lodge, D. (1984) *Small World,* London, Secker and Warburg.

Love, E. and Pimm, D. (1996) '"This is so": a text on texts', in Bishop, A. *et al.* (eds), *International Handbook of Mathematics Education,* Dordrecht, Kluwer Academic Publishers, pp. 371-409.

Mac Lane, S. (1986) *Mathematics: Form and Function,* New York, NY, Springer-Verlag.

MacHale, D. (1993) *Comic Sections,* Dublin, Boole Press.

Maher, P. (1994) 'Potential space and mathematical reality', in Ernest, P. (ed.), *Constructing Mathematical Knowledge: Epistemology and Mathematical Education,* London, Falmer Press, pp. 134-140.

Mandelbrot, B. (1989) 'Chaos, Bourbaki and Poincaré', *The Mathematical Intelligencer* **11**(3), 10-12.

Manin, Y. (1977) *A Course in Mathematical Logic,* New York, NY, Springer-Verlag, New York.

Mason, J. (2002) *Researching Your Own Practice: the Discipline of Noticing,* London, RoutledgeFalmer.

Matte Blanco, I. (1975) *The Unconscious as Infinite Sets: an Essay in Bi-Logic,* London, Duckworth.

May, K. (1972) 'Gauss, Carl Friedrich', in Gillespie, C. (ed.), *Dictionary of Scientific Biography,* New York, NY, Charles Scribner's Sons, pp. 298-315.

Mazur, B. (2003) *Imagining Numbers (Particularly the Square Root of Minus Fifteen),* New York, NY, Farrar, Straus and Giroux.

McAllister, J. (1996) *Beauty and Revolution in Science,* Ithaca, NY, Cornell University Press.

Menand, L. (2001) *The Metaphysical Club,* New York, NY, Farrar, Straus and Giroux

Menand, L. (2002) 'The principles of Oliver Wendell Holmes', in *American Studies,* New York, NY, Farrar, Straus and Giroux, pp. 31-53.

Menninger, K. (1958/1969) *Number Words and Number Symbols: a Cultural History of Numbers* (Broneer, P., trans.), Cambridge, MA, MIT Press.

Mighton, J. (1997) *Possible Worlds,* Toronto, ON, Playwrights Canada Press.

Millay, N. (ed.) (1956) *Collected Poems: Edna St. Vincent Millay,* New York, NY, Harper.

Mitchell, W. (1986) *Iconology: Image, Text, Ideology,* Chicago, IL, University of Chicago Press.

Molland, G. (1991) 'Implicit versus explicit geometrical methodologies: the case of construction', in Molland, G. (1995) *Mathematics and the Medieval Ancestry of Physics,* Brookfield, VT, Variorum, pp. 181-196.

Moon, F. (2003) 'Franz Reuleaux: contributions to 19th century kinematics and theory of machines', *Applied Mechanics Review* **56**(2), 261-285.

Mordant, I. (1993) 'Psychodynamics of mathematics texts', *For the Learning of Mathematics* **13**(1), 20-23.

Morgan, C. (1998) *Writing Mathematically: the Discourse of Investigation,* London, Falmer Press.

Morgan, F. (2005) 'Kepler's *conjecture* and Hales's proof', *Notices of the American Mathematics Society* **52**(1), 44-47.

Morrison, L. (1979) 'Poet as mathematician', in Robson, E. and Wimp, J. (eds), *Against Infinity: an Anthology of Contemporary Mathematical Poetry,* Parker Ford, PA, Primary Press, p. 45.

Movshovits-Hadar, N. (1988) 'School mathematics theorems: an endless source of surprise', *For the Learning of Mathematics* **8**(3), 34-39.

Mumford, D. (1991) 'A foreword for non-mathematicians', in Parikh, C., *The Unreal Life of Oscar Zariski,* Boston, MA, Academic Press, pp. xv-xxvii.

Mumford, D., Series, C. and White, D. (2002) *Indra's Pearls: the Vision of Felix Klein,* Cambridge, Cambridge University Press.

Murray, C. (2003) *Human Accomplishment: the Pursuit of Excellence in the Arts and Sciences, 800 B.C. to 1950,* New York, NY, HarperCollins.

Murty, R. (1988) 'Artin's conjecture for primitive roots', *The Mathematical Intelligencer* **10**(4), 59-67.

Nasar, S. (1998) *A Beautiful Mind: a Biography of John Forbes Nash, Jr.,* New York, NY, Simon and Schuster.

Nelsen, R. (1993) *Proofs without Words: Exercises in Visual Thinking,* Washington, DC, The Mathematical Association of America.

Nelson, S. (2002) 'Season for the naming of flowers', in *This Flesh These Words,* Victoria, BC, Ekstasis Editions, pp. 15-24.

Netz, R. (1998) 'Greek mathematical diagrams: their use and their meaning', *For the Learning of Mathematics* **18**(3), 33-39.

Netz, R. (1999) *The Shaping of Deduction in Greek Mathematics: a Study in Cognitive History,* Cambridge, Cambridge University Press.

Netz, R. (2000) 'Why did Greek mathematicians *publish* their analyses?', in Suppes, P. *et al.* (eds), *Ancient and Medieval Traditions in the Exact Sciences: Essays in Memory of Wilbur Knorr,* Stanford, CA, CSLI Publications, pp. 139-157.

Niven, I. (1947) 'A simple proof that π is irrational', *Bulletin of the American Mathematical Society* **53**(7), 509.

Norman, R. (1963) 'Whitehead and "Mathematicism"', in Kline, G. (ed.), *Alfred North Whitehead: Essays on His Philosophy,* Englewood Cliffs, NJ, Prentice-Hall, pp. 33-40.

Ord-Hume, A. (1977) *Perpetual Motion: the History of an Obsession,* New York, NY, St Martin's Press.

Osborne, H. (1984) 'Mathematical beauty and physical science', *The British Journal of Aesthetics* **24**(4), 291-300.

Padovan, R. (1999) *Proportion: Science, Philosophy, Architecture,* New York, NY, Routledge

Papadimitriou, C. (2003) *Turing: a Novel about Computation,* Cambridge, MA, MIT Press.

Pascal, B. (1659/2000) *Pensées,* Paris, Livre de Poche classique.

Peirce, B. (1850) 'Mathematical investigations of the fractions which occur in phyllotaxis', in *Proceedings of the American Association for the Advancement of Science* II, pp. 444-447.

Peirce, B. (1870/1882) *Linear Associative Algebra*, New York, NY, D. van Nostrand.

Penrose, R. (1974) 'The role of aesthetics in pure and applied mathematical research', *The Bulletin of the Institute of Mathematics and its Applications* **10**(7/8), 266-271.

Petsinis, T. (2000) *The French Mathematician,* New York, NY, Berkeley.

Picker, S. and Berry, J. (2000) 'Investigating pupils' images of mathematicians', *Educational Studies in Mathematics* **43**(1), 65-94.

Pickering, A. (1995) *The Mangle of Practice: Time, Agency, and Science,* Chicago, IL, University of Chicago Press.

Pimm, D. (1994) 'Another psychology of mathematics education', in Ernest, P. (ed.), *Constructing Mathematical Knowledge: Epistemology and Mathematical Education,* London, Falmer Press, pp. 111-124.

Pimm, D. (1995) *Symbols and Meaning in School Mathematics,* London, Routledge.

Pimm, D. (2001) 'Some notes on Theo van Doesburg (1883–1931) and his *Arithmetic Composition I*', *For the Learning of Mathematics* **21**(2), 31-36.

Pimm, D. (2004) 'A case of you: remembering David Fowler', *For the Learning of Mathematics* **24**(2), 16-17.

Pinker, S. (2002) *The Blank Slate: the Modern Denial of Human Nature,* New York, NY, Basic Books.

Plath, S. (1965) *Ariel,* London, Faber and Faber.

Plato *Philebus* (Waterfield, A., trans.), Harmondsworth, Middx, Penguin, 1982.

Plato *Symposium* (Rowe, C., ed. and trans.), Warminster, Wilts, Aris and Phillips, 1998.

Plaut, A. (1966) 'Reflections about not being able to imagine', *Journal of Analytical Psychology* **11**(2), 113-133.

Plotinus (204–270) *The Enneads* (Mackenna, S., trans.), New York, NY, Penguin, 1991.

Poincaré, H. (1908) *Science et Méthode,* Paris, Flammarion.

Poincaré, H. (1908/1956) 'Mathematical creation', in Newman, J. (ed.), *The World of Mathematics,* New York, NY, Simon and Schuster, vol. 4, pp. 2041-2050.

Polanyi, M. (1958) *Personal Knowledge: Towards a Post-Critical Philosophy,* New York, NY, Harper and Row.

Pólya, G. (1957, 2nd edn) *How to Solve It: a New Aspect of Mathematical Method,* New York, NY, Anchor.

Pólya, G. (1981) *Mathematical Discovery: on Understanding, Learning, and Teaching Problem Solving,* Vol. II, New York, NY, John Wiley and Sons.

Popper, K. (1972) *Objective Knowledge: an Evolutionary Approach,* Oxford, Clarendon Press.

Postrel, V. (2003) *The Substance of Style: How the Rise of Aesthetic Value Is Remaking Commerce, Culture and Consciousness,* New York, NY, HarperCollins.

Powell, A. and Frankenstein, M. (eds) (1997) *Ethnomathematics: Challenging Eurocentrism in Mathematics Education,* New York, NY, State University of New York Press.

Powell, A. and Frankenstein, M. (2001) '*In memoriam* Dirk Jan Struik: Marxist mathematician, historian and educator (30 September, 1894 – 21 October, 2000)', *For the Learning of Mathematics* **21**(1), 40-43.

Price, D. (1964) 'Automata and the origins of mechanism and mechanistic philosophy', *Technology and Culture* **5**(1), 9-23.

Radden, J. (2000) *The Nature of Melancholy: from Aristotle to Kristeva,* New York, NY, Oxford University Press.

Ramachandran, V. (2003) "The Artful Brain", Lecture 3 in the British Broadcasting Corporation's 2003 Reith Lectures, *The Emerging Brain.* (www.bbc.co.uk/radio4/reith2003/lecture3.shtml)

Read, H. (1960) *The Forms of Things Unknown: Essays towards an Aesthetic Philosophy,* New York, NY, Horizon Press.

Rebman, K. (1979) 'The pigeonhole principle', *The College Mathematics Journal* **10**(1), 3-13.

Rée, J. (1999) *I See a Voice: a Philosophical History of Language, Deafness and the Senses,* London, HarperCollins.

Regis, E. (1986) *Who Got Einstein's Office? Eccentricity and Genius at the Institute for Advanced Study,* Reading, MA, Addison-Wesley.

Reid, C. (1986) *Hilbert–Courant,* New York, NY, Springer-Verlag.

Resnik, M. (1980) *Frege and the Philosophy of Mathematics,* Ithaca, NY, Cornell University Press.

Reuleaux, F. (1876/1963) *The Kinematics of Machinery: Outlines of a Theory of Machines* (Kennedy, A., trans. and ed.), New York, NY, Dover.

Rice, A. (1999) 'What makes a great mathematics teacher?', *The American Mathematical Monthly* **106**(6), 534-552.

Richardson, J. (2001) *Vectors: Aphorisms & Ten-Second Essays,* Keene, NY, Ausable Press.

Riskin, J. (2002) *Science in the Age of Sensibility: the Sentimental Empiricists of the French Enlightenment,* Chicago, IL, University of Chicago Press.

Robinson, S. (2001) 'Why mathematicians now care about their hat color', *The New York Times,* April 10th, p. F5.

Rosenberg, H. (1964) *The Anxious Object: Art Today and its Audience,* New York, NY, Horizon Press.

Rota, G.-C. (1997) *Indiscrete Thoughts,* Boston, MA, Birkhäuser.

Rotman, B. (1987) *Signifying Nothing: the Semiotics of Zero,* London, Macmillan.

Rotman, B. (1988) 'Towards a semiotics of mathematics', *Semiotica* **72**(1/2), 1-35.

Rotman, B. (1993) *Ad Infinitum ...: the Ghost in Turing's Machine,* Stanford, CA, Stanford University Press.

Rotman, B. (2000) *Mathematics as Sign: Writing, Imagining, Counting,* Stanford, CA, Stanford University Press.

Rudd, T. (1651) *Elements of Geometry: the First VI Books, Whereunto Is Added the Mathematical Preface of John Dee,* London, Robert and William Leybourn for Richard Tomlins and Robert Boydell.

Russell, B. (1917) *Mysticism and Logic,* Garden City, NY, Doubleday Anchor Books.

Russell, B. (1956) 'Reflections on my eightieth birthday', in *Portraits from Memory and Other Essays,* London, G. Allen and Unwin, pp. 53-57.

Sabbagh, K. (2002) *The Riemann Hypothesis: the Greatest Unsolved Problem in Mathematics,* New York, NY, Farrar, Strauss and Giroux.

Santayana, G. (1896/1910) *The Sense of Beauty: Being the Outlines of Aesthetic Theory,* New York, NY, Charles Scribner's Sons.

Santayana, G. (1944) *Persons and Places: the Background of My Life,* vol. 1, New York, NY, Charles Scribner's Sons.

Sawyer, W. (1943) *Mathematician's Delight,* Harmondsworth, Middx, Pelican.

Sawyer, W. (1955) *Prelude to Mathematics,* Harmondsworth, Middx, Pelican.

Sawyer, W. (1970) *The Search for Pattern,* Harmondsworth, Middx, Pelican.

Scarry, E. (1999) *On Beauty and Being Just,* Princeton, NJ, Princeton University Press.

Schaaf, W. (ed.) (1948) *Mathematics: Our Great Heritage* (essays on the nature and significance of mathematics), New York, NY, Harper and Brothers.

Schapiro, M. (1946/1978) 'On a painting of van Gogh', in *Modern Art: 19th & 20th Centuries,* New York, NY, George Braziller, pp. 87-99.

Schattschneider, D. (1991) 'Counting it twice', *The College Mathematics Journal* **22**(3), 203-211.

Schattschneider, D. and Dolbilin, N. (1998) 'One corona is enough for the Euclidean plane', in Patera, J. (ed.), *Quasicrystals and Discrete Geometry, Fields Institute Monographs* (Vol. 10), Providence, RI, The American Mathematical Society, pp. 207-246. (See also 'Catalog of iso-hedral tilings by symmetric polygonal tiles' (mathforum.org/dynamic/one-corona), an interactive web site with Java Sketchpad versions of the tilings mentioned in Chapter 2.)

Schattschneider, D. and Emmer, M. (eds) (2003) *M. C. Escher's Legacy: a Centennial Celebration,* New York, NY, Springer-Verlag.

Schattschneider, D. and Fetter, A. (1991) *The Platonic Solids* (video and activity book), Berkeley, CA, Key Curriculum Press.

Schattschneider, D. and King, J. (1997) 'Preface: making geometry dynamic', in Schattschneider, D. and King, J. (eds), *Geometry Turned On: Dynamic Software in Learning, Teaching, and Research,* Washington, DC, The Mathematical Association of America, pp. ix-xiv.

Schechter, B. (2000) *My Brain is Open: the Mathematical Journeys of Paul Erdös,* New York, NY, Touchstone.

Scher, D. (2002, 2nd edn) *Exploring Conic Sections with The Geometer's Sketchpad,* Emeryville, CA, Key Curriculum Press.

Scher, D. (2003) 'What to expect when geometry becomes interactive?', *The New England Mathematics Journal* **35**(2), 36-43.

Schiralli, M. (1999) *Constructive Postmodernism: toward Renewal in Cultural and Literary Studies,* Westport, CT, Bergin and Garvey.

Schiralli, M. and Sinclair, N. (2003) 'A constructive response to *Where Mathematics Comes from*', *Educational Studies in Mathematics* **52**(1), 79-91.

Schogt, P. (2000) *The Wild Numbers,* New York, NY, Four Walls Eight Windows.

Searle, J. (1969) *Speech Acts: an Essay in the Philosophy of Language,* Cambridge, Cambridge University Press.

Seife, C. (2000) *Zero: the Biography of a Dangerous Idea,* New York, NY, Penguin.

Senechal, M. (1998) 'The continuing silence of Bourbaki: an interview with Pierre Cartier', *The Mathematical Intelligencer* **20**(1), 22-28.

Sfard, A. (1994) 'Reification as the birth of metaphor', *For the Learning of Mathematics* **14**(1), 44-55.

Shapin, S. and Schaffer, S. (1985) *Leviathan and the Air-Pump: Hobbes, Boyle, and the Experimental Life,* Princeton, NJ, Princeton University Press.

Silver, E. and Metzger, W. (1989) 'Aesthetic influences on expert math-ematical problem solving', in McLeod, D. and Adams, V. (eds), *Affect and Mathematical Problem Solving,* New York, NY, Springer-Verlag, pp. 59-74.

Sinclair, N. (2001) 'The aesthetic *is* relevant', *For the Learning of Mathematics* **21**(2), 25-32.

Sinclair, N. (2002) *Mindful of Beauty: the Roles of the Aesthetic in the Learning and Doing of Mathematics,* Unpublished doctoral dissertation, Kingston, ON, Queen's University.

Singh, S. (1998) *Fermat's Last Theorem,* London, Fourth Estate.

Sipser, M. (1986) The Reporter Column, *Technology Review* **89**(1), 80.

Skelton, R. (1993) 'Bion and problem solving', *For the Learning of Mathematics* **13**(1), 39-42.

Skemp, R. (1979) *Intelligence, Learning and Action: a Foundation for Theory and Practice in Education,* Chichester, Hants, Wiley.

Snow, C. (1959) *The Two Cultures and the Scientific Revolution* (The Rede Lectures), New York, NY, Cambridge University Press.

Solomon, Y. and O'Neill, J. (1998) 'Mathematics and narrative', *Language and Education* **12**(3), 210-221.

Spencer, J. (2001) 'Opinion', *Notices of the American Mathematical Society* **48**(2), 165.

St Augustine of Hippo (354–430), *De Genesi ad litteram* (Taylor, J., trans.), New York, NY, Newman Press, 1982.

St Theodore the Studite (759–826) *On the Holy Icons* (Roth, C., trans.), Crestwood, NY, St Vladimir's Seminary Press, 1981.

Stanley, D. (2002) 'A response to Nunokawa's article "Surprises in mathematics lessons"', *For the Learning of Mathematics* **22**(1), 15-16.

Stein, S. (1979) 'Existence out of chaos', in Honsberger, R. (ed.), *Mathematical Plums,* Washington, DC, The Mathematical Association of America, pp. 62-93.

Stein, S. (2001) *How the Other Half Thinks: Adventures in Mathematical Reasoning,* New York, NY, McGraw-Hill.

Steinberg, R. (1944) 'Solution to problem 4065', *The American Mathematical Monthly* **51**(2), 169-171.

Steiner, G. (2001) *Grammars of Creation,* New Haven, CT, Yale University Press.

Steiner, M. (1998) *The Applicability of Mathematics as a Philosophical Problem,* Cambridge, MA, Harvard University Press.

Steiner, R. (1992) *Toward a Grammar of Abstraction: Modernity, Wittgenstein and the Paintings of Jackson Pollock,* University Park, PA, Pennsylvania State University Press.

Stengle, S. (2000) 'An iconography of reason and roses', in Sarhangi, R. (ed.), *Bridges: Mathematical Connections in Art, Music, and Science* (Proceedings of the 2000 Conference), Winfield, KS, Southwestern College, pp. 161-168.

Stokes, A. (1965) *The Invitation in Art,* London, Tavistock Publications.

Stoppard, T. (1993) *Arcadia,* London, Faber and Faber.

Struik, D. (1998/2003) 'Foreword by Dirk J. Struik [1894-2000]', in Gerdes, P., *Awakening of Geometrical Thought in Early Culture,* Minneapolis, MN, MEP Publications, pp. vii-xi.

Sullivan, J. (1925/1956) 'Mathematics as an art', in Newman, J. (ed.), *The World of Mathematics,* New York, NY, Simon and Schuster, vol. 3, pp. 2015-2021.

Suri, M. (2002) *The Death of Vishnu,* New York, NY, Perennial.

Swetz, F. and Kao, T. (1980) *Was Pythagoras Chinese? An Examination of Right Triangle Theory in Ancient China,* Reston, VA, The National Council of Teachers of Mathematics.

Swift, J. (1745/1995) *Directions to Servants,* Harmondsworth, Middx, Penguin.

Sylvester, J. (1893) *The Educational Times* **46** (New Series, no 383), March 1st, p. 156.

Tahta, D. (1980) 'About geometry', *For the Learning of Mathematics* **1**(1), 2-9.

Tahta, D. (1991) 'Understanding and desire', in Pimm, D. and Love, E. (eds), *Teaching and Learning School Mathematics,* London, Hodder and Stoughton, pp. 220-246.

Tahta, D. (1994) 'On interpretation', in Ernest, P. (ed.), *Constructing Mathematical Knowledge: Epistemology and Mathematical Education,* London, Falmer Press, pp. 125-133.

Tahta, D. (1996) 'Mind, matter, and mathematics', *For the Learning of Mathematics* **16**(3), 17-21.

Tahta, D. and Pimm, D. (2001) 'Seeing voices', *For the Learning of Mathematics* **21**(2), 20-25.

Tamen, M. (2001) *Friends of Interpretable Objects,* Cambridge, MA, Harvard University Press.

Tasic, V. (1998) *Herbarium of Souls,* Fredericton, NB, Broken Jaw Press.

Tasić, V. (2001) *Mathematics and the Roots of Postmodern Thought,* New York, NY, Oxford University Press.

Taylor, R., Micolich, A. and Jones, D. (1999) 'Fractal analysis of Pollock's drip paintings', *Nature* **399** (6735), 422.

Thom, R. (1971) '"Modern" mathematics: an educational and philosophic error?', *American Scientist* **59**(6), 695-699.

Thom, R. (1973) 'Modern mathematics: does it exist?', in Howson, G. (ed.), *Developments in Mathematical Education,* Cambridge, Cambridge University Press, pp. 194-209.

Thom, R. (1975) *Structural Stability and Morphogenesis: an Outline of General Theory of Models* (Fowler, D., trans.), Reading, MA, W. A. Benjamin.

Thompson, D. (1917/1968, 2nd edn) *On Growth and Form,* 2 vols, Cambridge, Cambridge University Press.

Thurston, W. (1994) 'On proof and progress in mathematics', *Bulletin (New Series) of the American Mathematical Society* **30**(2), 161-177. (Reprinted in *For the Learning of Mathematics* **15**(1), 29-37.)

Thurston, W. (1997) *Three-Dimensional Geometry and Topology* (Vol. 1), Princeton, NJ, Princeton University Press.

Truss, L. (2004) *Eats, Shoots and Leaves: the Zero Tolerance Approach to Punctuation,* New York, NY, Gotham Books.

Turkle, S. (1978) *Psychoanalytic Politics: Freud's French Revolution,* New York, NY, Basic Books.

Turkle, S. and Papert, S. (1992) 'Epistemological pluralism and the revaluation of the concrete', *Journal of Mathematical Behavior* **11**(1), 3-33.

Tymoczko, T. (1993) 'Value judgments in mathematics: can we treat mathematics as an art?', in White, A. (ed.), *Essays in Humanistic Mathematics,* Washington, DC, The Mathematical Association of America, pp. 67-77.

Ullmo, J. (1962/1971) 'The geometric intelligence and the subtle intelligence', in Le Lionnais, F. (ed.), *Great Currents of Mathematical Thought,* vol. 2, New York, NY, Dover, pp. 5-13.

Unguru, S. (1994) 'Is mathematics ahistorical? An attempt to an answer motivated by Greek mathematics', in Gavroglu, K., Christianidis, J. and Nicolaidis, E. (eds), *Trends in the Historiography of Science,* Dordrecht, Kluwer Academic Publishers, pp. 203-219.

Upitis, R., Phillips, E. and Higginson, W. (1997) *Creative Mathematics: Exploring Children's Understanding,* London, Routledge.

Valens, E. (1964) *The Number of Things: Pythagoras, Geometry and Humming Strings,* New York, NY, Dutton.

van Dalen, D. (1981) *Brouwer's Cambridge Lectures on Intuitionism,* Cambridge, Cambridge University Press.

van der Waerden, B. (1930/1991) *Modern Algebra,* New York, NY, F. Ungar.

van Doesburg, T. (1930/1974) 'Comments on the basis of concrete painting', in Baljeu, J., *Theo van Doesburg,* New York, NY, Macmillan, pp. 181-182.

Vargas Llosa, M. (1999) *The Notebooks of Don Rigoberto,* New York, NY, Penguin.

Voland, E. and Grammer, K. (eds) (2003) *Evolutionary Aesthetics,* Heidelberg, Springer-Verlag.

von Fritz, K. (1975) 'Pythagoras of Samos', in Gillespie, C. (ed.), *Dictionary of Scientific Biography,* New York, NY, Charles Scribner's Sons, pp. 219-225.

von Neumann, J. (1947) 'The mathematician', in Heywood, R. (ed.), *The Works of the Mind,* Chicago, IL, University of Chicago Press, pp. 180-196.

Walkerdine, V. (1988) *The Mastery of Reason: Cognitive Development and the Production of Rationality,* New York, NY, Routledge.

Walter, M. (2001) 'Looking at a painting with a mathematical eye', *For the Learning of Mathematics* **21**(2), 26-30.

Warwick, A. (2003) *Masters of Theory: Cambridge and the Rise of Mathematical Physics,* Chicago, IL, University of Chicago Press.

Webster (1993, 10th edn) *Merriam-Webster's Collegiate Dictionary,* Springfield, MA, Merriam-Webster.

Wechsler, J. (ed.) (1978) *On Aesthetics in Science,* Cambridge, MA, MIT Press.

Weil, A. (1948) 'L'avenir des mathématiques', in Le Lionnais, F. (ed.), *Les grands Courants de la Pensée mathématique,* Marseille, Cahiers du Sud, pp. 307-320.

Weil, A. (1984) *Number Theory: an Approach through History from Hammurapi to Legendre,* Boston, MA, Birkhäuser.

Weil, A. (1992) *The Apprenticeship of a Mathematician* (Gage, J., trans.), Berlin, Birkhäuser.

Weil, S. (1952) *Gravity and Grace,* London, Routledge and Kegan Paul.

Weiss, G. and Haber, H. (eds) (1999) *Perspectives on Embodiment: the Intersections of Nature and Culture,* London, Routledge.

Wells, D. (1990) 'Are these the most beautiful?', *The Mathematical Intelligencer* **12**(3), 37-41.

Weyl, H. (1949) *Philosophy of Mathematics and Natural Science,* Princeton, NJ, Princeton University Press.

Weyl, H. (1952) *Symmetry,* Princeton, NJ, Princeton University Press.

Wheeler, D. (1979/2001) 'Mathematisation as a pedagogic tool', *For the Learning of Mathematics* **21**(2), 50-53.

Wheeler, D. (1982) 'Mathematization matters', *For the Learning of Mathematics* **3**(1), 45-47.

Whitehead, A. (1926) *Science and the Modern World,* New York, NY, Macmillan.

Whitehead, A. (1938) *Modes of Thought,* Cambridge, Cambridge University Press.

Whitehead, A. (1948) *Science and Philosophy,* New York, NY, Wisdom Library.

Whittaker, E. (1945) 'George David Birkhoff', *Journal of the London Mathematical Society* **20**(2), 121-128.

Wiener, N. (1956) *I Am a Mathematician: the Later Life of a Prodigy,* Garden City, NY, Doubleday.

Wigner, E. (1960) 'The unreasonable effectiveness of mathematics in the natural sciences', *Communications on Pure and Applied Mathematics* **13**(1), 1-14.

Wilson, E. (1978) *On Human Nature,* Cambridge, MA, Harvard University Press.

Wilson, E. (1998) *Consilience: the Unity of Knowledge,* New York, NY, Knopf.

Wilson, F. (1998) *The Hand: How Its Use Shapes the Brain, Language, and Human Culture,* New York, NY, Vintage.

Winnicott, D. (1967) 'The location of cultural experience', *The International Journal of Psycho-Analysis* **48**(3), 368-372.

Winnicott, D. (1971) *Playing and Reality,* London, Tavistock Publications.

Wittgenstein, L. (1922/1958) *Tractatus Logico-Philosophicus,* London, Kegan Paul, Trench, Trubner and Co.

Wittgenstein, L. (1953/1963, 2nd edn) *Philosophical Investigations* (Anscombe, G., trans.), Oxford, Blackwell.

Wolfe, T. (1975/2002) 'The painted word', in *Radical Chic and Mau-Mauing the Flak Catchers & The Painted Word,* London, Picador, pp. 1-99.

Woolf, V. (1931/1950) *The Waves,* London, Hogarth Press.

Woolfe, S. (1997) *Leaning Towards Infinity: How My Mother's Apron Unfolds into My Life,* London, Faber and Faber.

Yates, F. (1969) *Theatre of the World,* Chicago, IL, University of Chicago Press.

Yates, F. (1979) *The Occult Philosophy in the Elizabethan Age,* London, Routledge and Kegan Paul.

INDEX OF NAMES

INDEX

Printed in China